The Economics of Innovation

The Economics of Innovation

An Introduction

G. M. Peter Swann OBE

Professor of Industrial Economics, University of Nottingham and Director of Innovative Economics Ltd, UK

Edward Elgar
Cheltenham, UK • Northampton, MA, USA

Published by
Edward Elgar Publishing Limited
The Lypiatts
15 Lansdown Road
Cheltenham
Glos GL50 2JA
UK

Edward Elgar Publishing, Inc.
William Pratt House
9 Dewey Court
Northampton
Massachusetts 01060
USA

Paperback edition reprinted 2013, 2016

A catalogue record for this book
is available from the British Library

Library of Congress Control Number: 2008942911

ISBN   978 1 84844 006 7 (cased)
       978 1 84844 027 2 (paperback)

Printed and bound in Great Britain by the CPI Group (UK) Ltd

Contents

PART V   THE EFFECTS OF INNOVATION

PART VI   INNOVATION AND GOVERNMENT

List of figures

List of tables

Preface

This book stems from a second year undergraduate course on the *Economics of Innovation* at Nottingham University Business School. That course is an introductory one: all students have studied basic first year micro-economics, but have not studied the economics of innovation *per se*. However, the book is not only aimed at undergraduate students. It is designed for any reader with a basic knowledge of economics who wants to understand the role of innovation in economics.

Innovation is one of the most important economic and business phenomena of our time and is a topic of great practical and policy interest. It is a topic that still receives inadequate attention on most economics and business programmes. It is also a topic that is inadequately understood by economists and most others as well, and yet it has very widespread implications for our economy and society. I hope that those who read this book will learn something about what economics has to say about innovation and what economists can learn by studying innovation.

I wrote the book because none of the existing books on this topic, excellent though they are, could provide the sort of introductory text needed for this sort of course. Amongst these existing books, some of them are too advanced and/or technical for this level, some are too narrow in focus, and some concentrate on a few areas where there is a substantial literature rather than take an overview of the whole topic.

My objective in writing the book was to introduce the student to a broad range of issues around the economics of innovation in no more than 300 pages. Several friends and colleagues kindly provided very useful comments on a first draft. Two common comments were as follows. First, while the book covered many of the main topics, this was often done in a rather superficial way and important areas of the literature on each topic were overlooked. Second, the book contained some rather unexpected topics that my colleagues consider to be outside economics.

Thinking over these comments, and given my self-imposed page limit, it seemed that the only way of dealing with these criticisms would be to write a version with less breadth but more depth. But I was reluctant to go that way for one very important reason. An introduction of this sort should give the student an idea of many of the innovation issues (s)he should have on his/her 'radar screen'. Some of the students taking this course will not study innovation again, so I wanted to be sure they had some limited grasp of what I consider to be essential themes in the economics of innovation. And even those students destined for further study in the economics of innovation, need a broad introduction at this stage. For the path along which most economics courses tend to evolve is one of constant breadth but ever-greater depth: we look at the same issues, but in ever-greater technical detail. If the student does not get a broad introduction at the start, (s)he will not get it later.

For this reason, the book is unashamedly broad in scope and in places rather shallow. The book contains little technical detail on economic models. My criterion for choosing topics for inclusion is that these should be the most important issues from an empirical point of view, while the extent and sophistication of the literature is not one of my main criteria. At the end of the book, however, there is an appendix listing various supplementary readings to which the student and tutor may want to refer.

The analytical approach taken here is something of a hybrid: some mainstream economics, some evolutionary economics (and other 'heterodox' economics), a little history of economic thought and some ideas from adjacent disciplines (such as engineering, psychology and sociology). Why a hybrid approach? Because a hybrid approach is *essential* to understanding the economics of innovation. Pure economics is not enough!

The economics of innovation is one of the liveliest fields of industrial economics, relevant to business and policy, and one where many undergraduate students will find that they already have a lot of useful practical knowledge about innovation gained from everyday life.

Note on Readings and References

At the end of the book, the reader will find an appendix containing a list of suggested supplementary readings. This is followed by a complete list of those references actually cited in the main text. Those tutors and students looking for further material on particular topics will probably find that the first of these (the appendix) is the more useful. In addition, these supplementary readings will guide the students thorough various topics that are not covered in the present book.

Acknowledgements

I have studied the economics of innovation for over thirty years, since the beginning of my PhD. My ideas about the economics of innovation have been guided and influenced by many people, especially the following (but let me stress that none of them has any responsibility for errors or eccentricities in what follows): Cristiano Antonelli, David Audretsch, Rui Baptista, John Barber, Giuliana Battisti, Catherine Beaudry, Martin Binks, Daniel Birke, Alistair Bruce, Mario Calderini, Hugh Cameron, Bo Carlsson, Martin Cave, Gary Cook, Robin Cowan, Paul David, Giovanni Dosi, Kevin Dowd, Angela Dumas, Peter Earl, Walter Elkan, Dominique Foray, Chris Freeman, Luke Georghiou, Paul Geroski, Yundan Gong, Terence Gorman, Peter Grindley, Jonathan Haskell, Mike Hobday, Robert Hoffman, Tony Horsley, Effie Kesidou, Ray Lambert, Don Lamberton, Wassily Leontief, Franco Malerba, Silvia Massini, Stan Metcalfe, Jonathan Michie, Michio Morishima, Richard Nelson, Ray Oakey, Naresh Pandit, Keith Pavitt, Martha Prevezer, Tudor Rickards, Bob Rothschild, Mark Shurmer, Peter Skott, Luc Soete, Ed Steinmueller, Bob Stone, Paul Stoneman, David Stout, Paul Temple, Bruce Tether, Morris Teubal, Steve Thompson, Nick von Tunzelmann, James Utterback, Tony Watson, Richard Whitley, Paul Windrum and Vasilis Zervos. I am also grateful to Daniel Birke, Effie Kesidou, Cassey Lee Hong Kim, Qing-Ping Ma and Mounia Berrada-Gouzi for pointing out various errors in an earlier draft of the book, and to many students at Nottingham and Manchester Universities for their comments on things that were unclear in earlier versions of specific chapters.

I would like to express special thanks to Edward Elgar Publishing for their interest in my work, to Tim Baker for all his support, and to Jenny, Carri and Emma Swann for their constant encouragement.

PART I

Context

1. What is the economics of innovation about?

Some of the students approaching this book for the first time may well be thinking: 'On top of everything else, why do I need to study the economics of innovation?' Is there anything of particular interest that economists can learn from studying innovation, as opposed to all the other things we study? And is it not enough to learn some general principles of microeconomics which could be applied to issues of innovation as and when we need? Let me address these two points in turn.

I hope to demonstrate in this book that there are indeed many things of huge interest that economists can learn by studying innovation. Without doubt, innovation is one of the most important economic and business phenomena of our time. Innovation has very widespread implications for our economy and society but few of us understand these implications in full. Indeed, because innovation was largely ignored by mainstream economics until about thirty years ago, it is still true that innovation has a more important role within the economy than it does within the discipline of economics. But that is changing and the economics of innovation has become one of the very most popular areas of research for young economists.

I also hope to demonstrate that while the general principles of microeconomics do take us some way in understanding the economics of innovation, they are not sufficient. To develop a real understanding of the economics of innovation and a capacity to explore the many examples of innovation to be found in the real world, the student needs something more than standard microeconomics. (S)he needs at least some appreciation of evolutionary economics (and perhaps some other heterodox approaches to economics). (S)he also needs a little knowledge of the history of economic thought and a little knowledge of some adjacent disciplines (including engineering, psychology and sociology).

In short, therefore, I would say that the reasons to study the economics of innovation, on top of everything else, are that (a) innovation is incredibly important in the real economy and (b) the right way to study the economics of innovation is a bit different from the conventional economists' training.

THE FIELD

The economics of innovation has been concerned with five main groups of questions. First, how should we categorise and classify the different aspects of innovation? Second, how are innovations created? Third, how do customers react to innovations? Fourth, what effects do innovations have on the broader economy? And fifth, what can and should governments do to support and direct innovation activity? Parts II to VI of this book are organised around these five themes.

Aspects of Innovation

Part II (Chapters 3-8) will set out some of the key concepts used in defining, categorising and classifying innovations. Chapter 3 gives a broad overview of these various definitions and concepts and then Chapters 4-8 focus on specific issues: process innovation (Ch. 4); product and service innovation (Ch. 5); innovative pricing (Ch. 6); network effects (Ch. 7); and intellectual property (Ch. 8).

How Firms Achieve Innovation

Part III (Chapters 9-14) examines some of the essential steps in the making of innovations. This starts with a summary of various theories of creativity (Ch. 9), which originate outside economics, but about which the economist should have a basic understanding. We then turn to theories of the entrepreneur (Ch. 10). Although entrepreneurship and innovation are not the same thing, there is an important overlap between them. Chapter 11 describes how firms organise for innovation and shows that there are two leading models for such organisation depending on the type and source of the innovation. Chapter 12 explores the role of technology vision in organising for innovation. Then Chapters 13 and 14 address two phenomena which explain the macroeconomic organisation of innovation: industry clusters (Ch. 13) and the division of labour (Ch. 14).

Innovation and the Consumer

Part III (Chapters 15-16) examines the consumer response to innovation. While most of those reading this book will be familiar with just one neoclassical economic theory of the consumer, we shall show that a proper understanding of how customers react to innovation requires us to understand a broader range of theories of consumption. Chapter 15 describes six different theories of the consumer, starting with the traditional economic consumer but also including other theories from heterodox economics, sociology and anthropology. Chapter 16 draws out the connections between these theories of consumption and the diffusion of innovations, a topic of central interest in the economics of innovation.

The Effects of Innovation

In Part V (Chapters 17-21), our attention will focus on the effects of innovation on the broader economy. This can be analysed at various different levels. Chapter 17 looks at the implications of innovation for trade patterns. Chapter 18 examines the inter-relationship between innovation and market structure. This is a bi-directional relationship, since innovation changes market structure but market structure also influences the incentives for and scope for innovation. Chapter 19 looks at the role of innovation in wealth creation, and argues that the channels through which innovation can create wealth are both more numerous and more complex than is generally understood. Chapter 20 looks at the implications of innovation for competitiveness. We have separated the discussion of Chapters 19 and 20 because, while some may think that competitiveness and wealth creation are more or less the same thing, that is not really the case. Finally, Chapter 21 takes a brief look at the role of innovation in supporting a sustainable economy. This chapter will reveal two sides of innovation. It can sometimes help achieve sustainability but can also – perhaps unexpectedly – be a serious threat to sustainability.

Innovation and Government

Finally, in Part VI (Chapter 22), we look at whether government has a role in supporting and directing innovation. Chapter 22 argues that the past and present case for government involvement in innovation have been based on the argument that markets may not provide enough incentives for all innovation activities, and government has a role to correct this market failure. But beyond that, and beyond government's understandable wish to focus public resources on sectors where a country can be a serious competitor in the world market, there is no attempt on the part of government to decide on the direction of innovation. In a brief postscript to Chapter 22, we argue that government policy towards innovation will in future have to become more subtle if innovation is to support a sustainable economy and not make economic activity even less sustainable.

Context

History of Economic Thought

Before we embark on our survey of definitions and categories of innovation in Part II, however, Chapter 2 takes a rapid stroll through the history of economic thought, to get a flavour of what has been written about innovation. In that we shall see that many of the themes and issues to be covered in Parts II-VI have their roots in some quite old ideas.

2. Innovation in the history of economic thought

In this chapter we shall explore just a few of the things that economists have written about innovation since Adam Smith in 1776. In this short chapter, we cannot hope to do justice to all those scholars who have thought and written about innovation and the economy. But I hope that this chapter will convince the reader that it is always worth revisiting what earlier writers said about this subject even when – or perhaps, especially when – the perspectives of earlier writers are at odds with how we talk about the subject today.

Relatively few economics textbooks contain a discussion of the history of economic thought. Why is that? The usual explanation is a simple one. If an old scientific perspective is still valid and useful then it will be retained within the current state of the art. But if it is surpassed, then it should be forgotten. However, there are several reasons why a little history of economic thought helps to develop the student's understanding. For our present purpose, one reason is overwhelming. Study of the history of thought can often be the best way to get a subject back on track when it has strayed into an intellectual 'dead end'. When the neoclassical economics of the twentieth century seemed to have lost sight of the importance of innovation, other scholars revisited earlier writings to remind themselves of how and why innovation was seen to be so important in the eyes of earlier generations.

TWENTY ESSENTIAL INSIGHTS

In this section we follow a rapid tour through the history of economic thought about innovation, starting with Adam Smith. Not all these contributions are by people who we would now call economists, but all are key insights into the economics of innovation. This list obviously excludes many very important authors whose work will be cited later in the book. But my aim here is to try to bring out twenty different perspectives on the economics of innovation rather than to attempt any historically comprehensive survey.

I shall start with a chronological account and then turn to a thematic account.

Adam Smith (1723-1790) and John Rae (1796-1872)

A natural place to start is with Adam Smith who is often regarded as the founding father of economics. For Smith, invention and technological change were important factors in creating 'the wealth of nations'. But Smith reckoned that it was the division of labour rather than invention *per se* that was the main driving force in creating the wealth of nations. Invention itself followed from the division of labour (Smith, 1776/1904a, p. 11): 'the invention of all those machines by which labour is so much facilitated and abridged seems to have been originally owing to the division of labour.'

We discuss in Chapters 11 and 14 how, as Smith saw it, the division of labour would give rise to 'invention of all those machines'. But it is immediately interesting to contrast Smith's view with that of John Rae. Rae was and remains far less well known than Smith, but as argued by Brewer (1998), he was arguably the first to put invention at the centre of wealth creation. He was quite clear about this, writing of (Rae, 1834, Chapter 1): 'this power of invention, this necessary element in the production of the wealth of nations.'

Moreover, he took an opposite view to Smith on the relationship between invention and the division of labour (Rae, 1834, Appendix to Book 2): 'In [Smith], the division of labour is considered the great generator of invention and improvement ... In the view I have given it is represented as proceeding from the antecedent progress of invention.'

So, as Rae saw it, invention (rather than the division of labour itself) is the mechanism that lies at the heart of wealth creation. Chapter 14 contains further discussion of the contrasting perspectives of Smith and Rae.

John Stuart Mill (1806-1873)

The great philosopher and political economist John Stuart Mill also saw the centrality of invention in wealth creation (Mill, 1859/1929, p. 80): '... all good things which exist are the fruits of originality.' But he was one of the first to write of the paradox that invention did not obviously lead to an improvement in the lot of ordinary people (Mill, 1848/1923, p. 751):

> Hitherto [1848] it is questionable if all the mechanical inventions yet made have lightened the day's toil of any human being. They have enabled a greater population to live the same life of drudgery and imprisonment, and an increased number of manufacturers and others to make fortunes.

This paradoxical aspect of innovation is very important and will recur at various points throughout the book.

Karl Marx (1818-1883)

Karl Marx recognised the absolute centrality of innovation within economic development and a particular contribution was to identify the role of innovation in the competitive struggle. Writing in 1848, the same year as the last quotation from Mill, he observed (Marx and Engels, 1848): 'the bourgeoisie cannot exist without constantly revolutionising the instruments of production, and thereby the relations of production, and with them the whole relations of society.'

This is an essential insight. Marx was talking in particular of the effects of innovation on the class struggle but the point is a much more general one. Replace 'bourgeoisie' (which means in colloquial English, 'good burghers') by the name of a company in an ultra-competitive market and the meaning of this quotation is not far from the common saying, 'innovate or die'. Chapter 20 is concerned with the role of innovation in ensuring competitiveness.

John Ruskin (1819-1900)

Our next economist should, perhaps, not be given that title, because he is better known as a philosopher and historian of art. And yet, he himself considered that his writings on political economy were greater than any of his other writings. John Ruskin's writings are fascinating for the very different view that he held of invention, on the one hand, and innovation, on the other. His description of invention was almost always favourable. He wrote of: 'noble invention'; 'the vigour of its invention'; 'the majesty of invention'; and 'the rare gift of invention'. But his view of innovation was generally quite scathing. He wrote of: 'violent innovation'; 'restlessness of innovation'; 'crash of innovation'; 'revolting innovation'.[1]

To those brought up in the Schumpeterian tradition of innovation, this seems a peculiar difference. Granted, invention is distinct from invention in the Schumpeterian scheme, but the difference is one of timing and execution rather than a difference of kind. Schumpeterian invention is an essential precursor to innovation, and the latter only takes place when prior inventions are applied commercially. So why did Ruskin find a difference of kind? The answer is that Ruskin saw many other channels for the use of invention other than what we would now call innovation and he believed that these other channels were much more benign than innovation. To understand this unusual perspective requires a careful analysis of how invention and innovation create wealth, and we turn to that in Chapter 19.

Henry George (1839-1897)

The next economist in our chronology is an economist, for certain, but his inclusion may seem surprising because he is relatively unknown to modern economists. Henry George's most famous book *Progress and Poverty* eminently deserves inclusion here because it made such an important contribution to understanding what I have called the paradoxical aspect to innovation. Like Mill before him, George considered that innovation would not necessarily improve the lot of the ordinary person, though his reasoning was a bit different (George, 1879/1931, pp. 176-177):

> every improvement or invention, no matter what it be, which gives to labour the power of producing more wealth, causes an increased demand for land and its direct products ... This being the case, every labour-saving invention, whether it be a steam plough, a telegraph, an improved process of smelting ores, a perfecting printing press, or a sewing machine, has a tendency to increase rent.

George also said (1879/1931, p. 212): 'There are many persons who still retain a comfortable belief that material progress will ultimately extirpate poverty ... but the fallacy of these views has already been sufficiently shown.' While several subsequent writers have disputed the theoretical logic and the empirical accuracy of this assertion, it is still worthy of consideration. Today we find that property prices in strong business clusters continue to soar ahead of prices in the hinterland, despite the arguments in many quarters that cheap communications and transport will lead to a dispersion of economic activity and an equalisation of property prices. George was right to posit a relationship between innovation and property prices though the present mechanism is a little different from the one he saw.

Henry George's insight is important for this reason and because he saw one of the several paradoxes about innovation. Innovation does not necessarily do what you expect and/or may have unexpected side effects.

Alfred Marshall (1842-1924)

The next economist in our list is Alfred Marshall. The particular passage of Marshall I quote below (Marshall, 1920, p. 90) builds on an essential insight in McCulloch (1864/1965, p. 23):

> The gratification of a want or a desire is merely a step to some new pursuit. In every stage of his progress he is destined to contrive and invent, to engage in new undertakings; and when these are accomplished to enter with fresh energy upon others.

Marshall's consumer is, in a sense, an innovator. We shall see in Chapter 15 (on consumption) that this type of consumer behaviour stands in interesting distinction to other types of consumer behaviour. This is, in today's language, the idea that innovation is not just the prerogative of the producer but can be an action on the part of the consumer. Until very recently, many economists and policy makers had grave difficulties with this idea. But with the more recent work of von Hippel (2005), in which the customer (and even the consumer) has a role in innovation, it does not seem so strange.

Thorstein Veblen (1857-1929)

The next economist on our list could be classified as an economist and a sociologist. Thorstein Veblen had an essential insight into the issue of conspicuous consumption – the desire to consume so that the consumer could draw attention to him/herself. We shall encounter this Veblen consumer in Chapter 15.

For now, Veblen's essential insight was – in the eyes of some – a piece of mischief. He took the old proverb, 'necessity is the mother of invention' and turned it on its head (Veblen, 1914, p. 315): 'Invention is the mother of necessity.'

What did he mean by that? Whereas the original version of the proverb suggests that invention is driven by user need, Veblen was suggesting that invention had by then become driven by another need – the need of the innovator to create competitive distinction. An innovator who creates competitive distinction cannot necessarily be sure that his/her competitively distinct product actually meets a consumer need. But with the right consumers in place (what we shall call below, *Veblen consumers*), there will always be a demand for distinction. So Veblen was suggesting that a demand could emerge for inventions for which there was no original need on the part of the consumer.

We shall revisit this at length in Chapter 15. Marshall and Veblen consumers play important roles in the emergence of markets for innovations.

Joseph Schumpeter (1883-1950)

The next economist on our list is one of the most important – possibly the most important of all – in the history of economic thought about innovation. Joseph Schumpeter captured, perhaps better than anyone before or since, the dual role of innovation in economics. He described (Schumpeter, 1954, p. 83): '[the] process of industrial mutation ... that incessantly revolutionizes the economic structure from within, incessantly destroying the old one, incessantly creating a new one. This process of *Creative Destruction* is the essential fact about capitalism.' And he added (Schumpeter, 1954, p. 83-84): 'Every piece of business strategy acquires its true significance only against the background of that process and within the situation created by it. It must be seen in its role in the perennial gale of creative destruction.'

This concept of *creative destruction* is one of the most important in the economics of innovation. The innovator creates something – competitive advantage probably and possibly wealth – but in doing so destroys something else, often the competitive position of a rival firm. So innovation creates and destroys at the same time, but with luck the value of creation will exceed the value of destruction.

Schumpeter also emphasised a very important point which contrasts with the conventional view within neoclassical economics. He insisted that this creative destruction was a much more important force for competition that the traditional concept of price competition (Schumpeter, 1954, pp. 84-85):

> in capitalist reality as distinguished from its textbook picture, it is not that[2] kind of competition which counts but the competition from the new commodity, the new technology, the new source of supply, the new type of organisation ... competition which commands a decisive cost or quality advantage and which strikes not at the margins of the profits and the outputs of existing firms but at their foundations and their very lives. This kind of competition is as much more effective than the other as a bombardment is in comparison to forcing a door, and so much more important that it becomes a matter of comparative indifference whether competition in the ordinary sense functions more or less promptly.

Lewis Mumford (1895-1990)

The next on our list was not an economist as such but his writings on *technics* (as he called them) and society should be essential reading for any student of economics of innovation. Lewis Mumford made a most memorable observation about technology and the industrial revolution (1934, p. 14): 'The clock, not the steam-engine, is the key-machine of the modern industrial age.' Not all historians would agree with this assertion by any means, and some would insist that the steam-engine is the more important. But it is an interesting argument, because it is part of Mumford's persistent thesis that the most interesting innovations have been social rather than technological as such. Certainly, anyone who looks at the economic history of the clock will see what an extraordinarily profound influence it has had on economic development (Cipolla, 1967; Landes, 1983).

John Kenneth Galbraith (1908-2006)

John Kenneth Galbraith was a great populariser of economics. Because of this he was not always treated with great respect by academic economists, though he has been immensely influential amongst non-economists. Galbraith had some controversial views about the value of economic growth. One of his most famous sayings was the following (Galbraith, 1958, pp. 152-153): 'As a society becomes increasingly affluent, wants are increasingly created by the process by which they are satisfied.'

The implication of this, as Galbraith stressed in some detail, is that innovations may not really increase wealth in a true sense if the demands satisfied by an innovation were demands created by the innovator and marketer. We shall need to return to this point in Chapter 15, when we encounter Galbraith's perspective on consumption of new products (or innovations).

E.F. Schumacher (1911-1977)

The next on our list is best known as one of the pioneers of environmental economics. E.F. Schumacher is most famous for his book, *Small is Beautiful*, one of the pioneering books on the economics of sustainability. In this, he makes a memorable assertion about technological progress (Schumacher, 1974, p. 26): 'man is far too clever to be able to survive without wisdom.' By this he means that clever technological progress is providing us with extraordinary powerful tools to change the world, but all of these tools can have bad effects as well as good, and we cannot work out which will do good and which bad without a great deal of wisdom. Schumacher, amongst many others, believed that the same social processes that have made us clever (including the division of labour in science) have tended to reduce our wisdom.

Chris Freeman[3]

If I were asked to pick out a single living economist who has made the biggest contribution of all to the economics of innovation, I would choose Chris Freeman. He has been an energetic and charismatic pioneer for six decades. Perhaps his most important contribution is not just *what* he has done but *how* he has done it. Freeman's work offers the most compelling case for the view that economists will understand innovation best if they treat it as a multidisciplinary field. Or, to put it another way, economics on its own is not enough to understand the economics of innovation. This is a view that I have already espoused in the preface to this book and explains some of the unexpected items in this book. But in truth, the multidisciplinarity in this book is a mere shadow of what is to be found in Freeman's work.

As just one example, a recent short (17 page) article of his (Freeman, 2007) discussed some recent developments in information and communications technology from the perspective of *all* the following disciplines or fields: classics, design, development economics, journalism, law, linguistics, management, philosophy, policy, political economy, science and technology policy, sociology, strategy and technology management. And if we look over other work by Freeman, we find an even wider spread than that.

In the eighteenth and nineteenth centuries, such a command of a wide range of disciplines would not be exceptional. For example we find it in the work of three of those we have met above: Smith, Mill and Ruskin. But now it is a real rarity.

Kenneth Arrow (1921-)

Arrow has left his mark on almost every branch of economics and the economics of innovation is no exception. In a celebrated and frequently cited article of 1962, he demonstrated why we may experience market failure in the allocation of resources for invention (Arrow, 1962, p. 619):

> To sum up, we expect a free enterprise economy to underinvest in invention and research (as compared with an ideal) because it is risky, because the product can be appropriated only to a limited extent, and because of increasing returns in use. This underinvestment will be greater for more basic research. Further, to the extent that a firm succeeds in engrossing the economic value of its inventive activity, there will be an underutilization of that information as compared with an ideal allocation.

As we shall see in Chapter 22, these arguments underpin much of the current economic case for government policy towards invention and innovation.

Robert Solow (1924-)

Before the work of Robert Solow, mainstream economics treated technical change as one of the exogenous factors influencing economic activity, but not necessarily a factor of huge importance. But Solow (1957, p. 320) demonstrated otherwise: 'Gross output per man hour doubled over the interval [1909-1949], with 87½ per cent of the increase attributable to technical change and the remaining 12½ per cent to the increased use of capital.'

The general reaction of the economics profession to this result was one of great surprise. Before that, the general assumption would have been that increased use of capital would have been far more important than technical change. But here, Solow turned everything on its head, by asserting that technical change was far more important. This observation (along with some of the remarks of Schumpeter) was one of the things that caused economists to start taking innovation seriously again.

More recently, Solow made a famous observation about one of the paradoxes that surround innovation (Solow, 1987, p. 36): 'we see computers everywhere except in the productivity statistics.' This remark has been very widely quoted and is often called the IT productivity paradox. Draca et al. (2007) offer a recent review of the evidence on this IT productivity paradox.

Nathan Rosenberg (1927-)

Rosenberg is one of the most influential economic historians working on the economics of innovation. Amongst his many other contributions, Rosenberg did us a great service by making it clear that we should stop thinking about technological change as if it were *either* demand pull *or* technology push (Mowery and Rosenberg, 1979, p. 143):

> Rather than viewing either the existence of a market demand or the existence of a technological opportunity as each representing a sufficient condition for innovation to occur, one should consider them each as necessary, but not sufficient, for innovation to result; both must exist simultaneously.

This is not an easy passage and the reader may want to re-read it several times. But the message is very important. Innovation cannot be demand-pull or technology-push alone. We must have both working together before an innovation will take off.

Richard Nelson (1930-)

Richard Nelson is another of the most important economists working on the economics of innovation in the second half of the twentieth century. One of his most influential works is *An Evolutionary Theory of Economic Change* (Nelson and Winter, 1982). While it was not the first work to discuss the potential of using an evolutionary metaphor within the economics of innovation, this book did create the foundations for what has since become known as *evolutionary economics*. Building on the Schumpeterian perspective, in which innovation is a *perennial gale*, evolutionary economics uses the biological metaphor of evolution to understand how innovations and the economy co-evolve.

As well as writing of the power of innovation, Nelson and Winter also explained why innovation poses such a challenge to the innovating firm. Most organisations depend on routines in working life to avoid too much intra-organisational strife. The larger and more complex the organisation, the more important are these routines. They describe what each person does and give the standard by which their performance can be assessed. But as Nelson and Winter (1982, p. 128) put it, very simply: 'innovation involves change in routine.'

Innovation requires some or all employees to do things differently from how they have been done in the past. Some employees may be content with this but others will not be. For this reason the consequences of an innovation for an organisation may not become clear for some time, until all the consequences of routine change have been worked through.

Paul David (1935-)

Paul David is famous as one of the originators of the idea of *path dependence* in the economics of innovation. This, roughly speaking, is the idea that the equilibrium to which an economic process converges depends on the precise historical path by which it got there.[4] This stands in contrast to the traditional assumption in neoclassical economics that the equilibrium reached is not influenced by the path by which the economy reaches that equilibrium.

Path dependence is of importance and interest in various areas of the economics of innovation. We shall visit one of the most important in Chapter 7. There we see that the question of which technology will emerge as a standard as a new technology market grows will depend on details of history. Paul David gave us one of the most striking examples of this: the QWERTY keyboard – that is, the traditional layout of an English-language typewriter keyboard. David (1985) argued that but for some accidents of history, we could have ended up using a very different keyboard layout. Because, for sure, the QWERTY layout is not the best design from the viewpoint of ergonomic efficiency. As David (1985, p. 336) says: 'Competition in the absence of perfect futures markets drove the industry prematurely into standardisation on the wrong system.'

This perspective poses quite a challenge to the neoclassical economist who believes that competitive markets could not possibly converge on the wrong outcome. And some have disputed whether the QWERTY keyboard is really an example of lock-in to an inferior standard. Nevertheless, David is probably quite right to conjecture that 'there are many more QWERTY worlds lying out there' and definitely right to assert that 'history matters' in such processes. Any text on the economics of innovation must give proper attention to path dependence and the possibility of lock-in, and we shall do this in Chapter 7.

Eric von Hippel (1941-)

While not an economist in the narrow sense, Eric von Hippel is another very influential scholar of innovation. Perhaps his most influential contribution was to point out just how important the user, customer or consumer is to the process of innovation. Indeed, in more recent work, von Hippel moved on from this to talk about a view of innovation in which the user is not passive, nor just an advisor in the innovation process, but where the user is the centre of gravity in the innovation process (von Hippel, 2005, p.1):

> When I say that innovation is being democratized, I mean that users of products and services – both firms and individual consumers – are increasingly able to innovate for themselves. User-centered innovation processes offer great advantages over the manufacturer-centric innovation development systems

This follows on in a natural way from the Marshall/McCulloch view (described above) in which the consumer/customer/user can be the driver of innovation and not just a passive recipient.

Paul Geroski (1952-2005)

The last economist on my list is the late Paul Geroski – a very influential academic economist and, at the time of his death, Chairman of the UK Competition Commission. One of his last works re-examined the contrasting incentives for an innovator to be first to market or on the other hand to be what is generally called in the strategy literature, a 'fast second' (Geroski and Markides, 2004):

> Normally, it does not make much sense for a firm to move fast to be first into a new market. Two quite different considerations favour followers who play a 'fast second' strategy. First, there are 'time-cost trade-offs' which penalize firms that try to do things too quickly ... The second advantage that fast second movers have is that they can free ride on the efforts of first movers.

Geroski was not the first to make this observation that a fast second strategy was often more successful than a fast first strategy. But we have cited his discussion of the idea because, arguably, he did as much with the idea as anyone else – both as an academic and as one of the relatively few academics who has been very influential in the policy world.

THEMES

Those are our twenty perspectives on the economics of innovation. We can summarise these contributions under five thematic headings. This is important because these five themes are essential areas to revisit later in the book.

Innovation and Wealth Creation

The earliest perspectives, those of Adam Smith and John Rae, saw innovation as something that lies at the heart of wealth creation. This perspective recurs in the quotations from Mill, Solow (especially) and Rosenberg. And, of course, Schumpeter saw innovation as an essential force for wealth creation – though as part of a process of creative destruction and not just creation.

We shall look at the role of innovation in wealth creation in Chapter 19. We shall argue there that the traditional economic perspective is too narrow and in reality, innovation can contribute to wealth creation in many ways that have not been charted by economics.

Innovation and Competitiveness

A rather different view of innovation started with Marx. For him, innovation was an essential part of a competitive battle. It might also be an important force for wealth creation but if we were to look at innovation from that perspective alone we would not see the full picture. For, in innovation, the innovator's primary aim is to win a competitive battle. This perspective was developed much further in Schumpeter's concept of *creative destruction*. Moreover, in some of the other quotations above (notably Veblen and Geroski), innovation has a role that is distinct from simply increasing the well-being of customers.

We shall look at the role of innovation in promoting competitiveness in Chapter 20. We shall see there that the sorts of innovations that promote competitiveness and the sorts of innovations that increase wealth are not necessarily the same. That may be surprising to the reader: we hope that by the time (s)he has had read this book, this statement will not seem such a puzzle.

Innovation and Sustainability

A third perspective on innovation started with Ruskin and was developed especially by Schumacher. This is the question of whether innovation helps to achieve sustainability or, on the contrary, is partly responsible for the unsustainable trajectory on which economic development has embarked. We shall visit this in Chapter 21.

Innovation, Unexpected Side Effects and Paradoxical Non-Effects

Several quotations have noted that innovation sometimes has unexpected (and possibly undesirable) side effects – notably George, Veblen, Mumford,

Ruskin. Other quotations have argued that innovation sometimes has no effect when we would expect that it would have an effect – notably Mill, George, Galbraith, Solow, David. We shall see several examples of this in what follows. It is very important that those who study innovation maintain a careful look-out for these unexpected side effects and paradoxical non-effects. Too often, we have been persuaded by technologists and engineers that the effects of innovation will be obvious when in fact they are much more subtle.

Through which Channels Does Innovation Work?

The final perspective to stress here is the idea that the firm is not the only innovator in the economy. Careful reflection on many of the above quotations (notably Marshall, von Hippel, Veblen, and Mumford) will convince us that innovative activity is much more widespread, and by no means in the monopoly of the innovative producer. Moreover, we shall see that many of the effects of innovation pass through what may seem to be unexpected channels – see Chapter 19.

NOTES

[1] Various passages from Cook and Wedderburn (1903-1912/1996).
[2] By this, Schumpeter means the traditional neoclassical concept of price competition.
[3] I have not been able to find information on date of birth.
[4] A more precise definition is given by David (1997, p. 13-14).

PART II

Aspects of innovation

3. Basic concepts in innovation

The aim of this chapter is to give the reader a brief introduction to some of the most important and basic concepts in innovation. Many of these will be developed in more detail later in the book.

To get us started, we shall use a very simple model of innovation. This simple model is sometimes called the 'linear model'. In this context, the word 'linear' is not used in the conventional mathematical sense. It means that research and creativity lead to innovation and wealth creation in a 'straight line' (Figure 3.1) and that this is a one-way process.

We shall see later in the book that the relationship between creativity and wealth creation is not a straight line, but may follow many complex paths. Moreover, it is not a one-way process. Most notably, feedback from customers to innovators is one of the most important inputs to the innovation process. Nevertheless, a more realistic model is also a much more complex model and that will need to wait until later in the book (Chapter 19). For now, the simple model will help us get started, but the reader must bear in mind that it is a huge simplification.

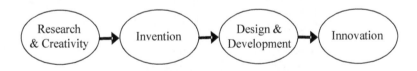

Figure 3.1   The simplistic linear model of innovation

With luck and hard work, research and creativity will generate inventions. But this is just the first step on the route to an innovation. The invention is just an idea. Before it can be turned into a commercially viable innovation, a lot of development and sometimes design work is required. Not all inventions will make it through this stage to become innovations. We reserve the word *innovation* for when it is first put to commercial use. That is, an innovation takes place when the new idea is first used in a company's production process or is first offered for sale in a market.

At each stage of this process, various issues crop up that we shall discuss in the book. These are summarised in Table 3.1 below. Along the top of this table, we list some of the main elements in the 'linear model'. Then in each column, we list some of the supplementary issues relevant to that stage. For example, in the column headed 'Research & Creativity', we list two different theories about the creative process. While strictly speaking these two theories derive from the literature on creativity and not the economics literature, it is useful for the student of economics to have a basic understanding of them. In the column headed 'Invention' we list some of the techniques companies use to protect the potential commercial value of their inventions. In the column headed 'Innovation & Technological Change', we list the different types of product innovation and the different types of process innovation. Finally, the column headed 'Market' lists two different theories of the consumer: passive and active. This chapter gives a very brief account of each of these supplementary issues, and then they are discussed in greater detail later in the book.

Table 3.1   Issues in the innovation chain

Research & Creativity	Invention	Innovation & Technological Change	Market
Autonomous Theory	IP Protection, Formal: Patent Registered Design Trademark Copyright	Innovations: Product Pricing Service Proliferation	Passive Consumers
Combinatorial Theory	IP Protection, Informal: Complexity Lead-Time Confidentiality Secrecy	Innovations: in Process in Organisation in Business Model in Supply Chain in Marketing	Active Consumers
	Open Innovation	Innovations: Incremental Radical Architectural	

The table gives the agenda for this chapter. We shall start with the main terms in the top row and then turn to the themes in each column.

THE MAIN LINKS IN THE LINEAR MODEL

In this section, we shall define and discuss the main terms used in the top row of Table 3.1. The one exception is that we assume it is not necessary to define what a market is, as all the students reading the book will have some background in economics.

What Is Innovation? How Does it Differ from Invention?

There seem to be a bewildering variety of definitions because the word is widely used and not just by economists. Within economics, however, innovation has come to have a fairly specific meaning. One very concise and popular definition of innovation is the following: 'The successful exploitation of new ideas'.[1]

This captures two of the essential features of innovation. It is not just about the generation of new ideas; it is about their *commercial exploitation*. This definition also helps us to understand the clear distinction made in economics between *innovation* and *invention*.

Invention is about the generation of new ideas, whether by research or other forms of creativity. *Inventions* are the culmination of research activity and are ideas, sketches or models for a new product or process, that may often be patented. But invention stops short of commercial use or exploitation. It is when the new idea is used in the market that we have *innovation*. Innovation is the commercial application of invention. Many inventions never turn into innovations, and for those that do there can be a long and complex chain of events between invention and innovation – rather more complex than our simple model suggests.

We should add here that the word *imitation* often crops up in the discussion of invention. If company B copies a successful innovation by company A, then we call that imitation. The time lag between innovation and imitation can be very variable, depending as it does on patent rights, lead-times and so on.

What Is Creativity? How Does it Differ from Invention?

Once again, it is hard to pin this down because the two words 'creativity' and 'invention' are sometimes used interchangeably. But within this book we treat 'creativity' as the *process* or *activity*, and 'invention' as the *result*. Creativity is a long and sometimes painful process. Some indeed would say it is better described as an 'activity' than a 'process' because the there are no rules for creativity, or if there are, nobody knows what they are. Even the greatest creative minds don't always know how their creative processes work. By contrast, an invention can be described on paper – and indeed will have to be if the inventor wishes to patent it.

Research and Development (R&D)

To some observers, putting *research* together with *development* is to group two rather different activities, and also tends to make us forget that in an industrial context most corporate spending on *R&D* should properly be described as *D* rather than *R*. *Basic research* produces new scientific knowledge, hypotheses and theories and these are expressed in research papers and memoranda, while inventive work drawing on this basic research produces patentable inventions. *Development* work, on the other hand, takes this stock of knowledge and patentable inventions as its raw materials and develops blueprints, specifications and samples for new and improved products and processes (Freeman, 1982). Basic research can be (some would say, *should* be) some way from the market, while development work is much nearer the market.

Design or Development?

The Design Council (1995) published a booklet with 50 different definitions of design offered by famous designers, business-people, politicians, and others. One of the most useful is by Michael Wolff (here quoted in abbreviated form): 'Design is a vision … Design is a process … Design is a result.' This three-faceted definition is very useful and many of the other definitions focus on one or other of these three facets. Another valuable definition (by John Harvey Jones) is the following: 'Design adds the extra dimension to any product.'[2]

Bernsen (1987) helps us to see design in its broader competitive context. He notes that manufacturing skills, engineering know-how or high-quality materials can no longer be a source of lasting competitive distinction, because industrial products are manufactured all over the world, using engineering know-how which is sold globally. But Bernsen goes on:

> What will make a product stand out is the quality of the way it matches the purpose, skills and personality of the user, of the visual communication which goes with it, of the environment in which it is sold, and of the image of its maker. All of these are created by design.

Design is not however the *only* route by which creativity is harnessed for commercial benefit. The core economics literature pays more attention to the role of development (see above) as a channel for creativity in the innovation process.

Is there a Difference between Innovation and Technological Change?

Finally, to complete our discussion of the terms in the top row of Table 3.1, we need to explain the difference between innovation and technological change. The early economics literature on innovation (pre-1980, say) tended to talk about the economics of technological change[3] rather than the economics of innovation. Is there any important difference here?

The difference is not of huge importance, but the main point the student needs to bear in mind is that innovation is a *wider* concept than technological change. Indeed, technological change is a sub-set of innovation. All technological changes are innovations, but not all innovations involve technological change. Some innovations may involve new packaging and new design but not any really new technology. The various generations of Sony 'Walkman' are a good example of this. As innovations these were very successful. But this was not because they embodied new technology. Indeed, it can be argued that the Walkman just used existing technology but packaged and designed the product in a most imaginative way.[4]

Throughout this book we shall focus on the wider concept of innovation. While several of the practical innovations discussed in the book involve technological change, several others do not.

DIFFERENT FORMS OF INNOVATION

Now we turn to the additional issues mentioned in Table 3.1. We start with the various types of innovation discussed in the column headed 'Innovation and Technological Change'.

Joseph Schumpeter (1954), perhaps the most influential writer about innovation, demonstrated that innovation could take many forms:

> introducing new commodities or qualitatively better versions of existing ones; finding new markets; new methods of production and distribution; or new sources of production for existing commodities; or introducing new forms of economic organisation.

Following on from this, economics has developed quite a detailed terminology to describe different types of innovation. Here we shall concentrate on just the most important elements of that terminology.

Product or Process Innovation?

In this book, I argue that the distinction between product and process innovation is a very important one. Not everyone agrees with that. Some have suggested that the distinction between product and process innovation is not a useful or important one because the same thing can be a product innovation to one person and a process innovation to another. Thus, they argue, a new improved computer may be a *product innovation* to the company selling it, but it is a *process innovation* to the operations manager using the new computer to run a more efficient production line.

While the example is correct in one sense, this does not mean the argument is correct. It is rather like saying there is no difference between a mother and a daughter, because the same person can be both a mother and a daughter. Clearly a mother is not the same as a daughter. In the same way, we shall see in Chapters 4 and 5 that process innovation is *not* the same as product innovation. The distinction is a vital one because these two types of innovation have very different economic effects.

A *pure process* innovation simply changes the way in which a product is made, without changing the product itself (except perhaps the price at which it will be sold). A *pure product* innovation, on the other hand, creates a new or improved product for sale without any change in the production process – except that more inputs (labour, machine time and materials) may be required. In practice, many if not most innovations embody some of each. Often a new and improved process will lead to incidental improvements in the product, and even more frequently a new product will require some innovations in the production process. Nevertheless, the distinction is an important one, as the next two chapters will show.

Different Types of Product Innovation

In economics, we use the *characteristics* approach to analyse and categorise different types of product innovation. This is described in more detail in Chapter 5, but the basic idea is as follows. We describe a product in terms of a list (sometimes a very long list) of features or characteristics. We can then compare different varieties of the same product in terms of their different scores on all these characteristics. This is useful because it allows us to make a distinction between product innovations that affect only one characteristic, product innovations that affect several characteristics, product innovations that introduce just one *new* characteristic[5] and product innovations that introduce so many new characteristics that we may wish to call the innovation a completely new product.

It is increasingly common for different firms to seek competitive distinction by innovations in the service they offer rather than innovations in the product *per se*. This growth in the service element is partly a consequence of increasing specialism and division of labour (Chapter 14). Certainly in economies such as the UK, a much larger share of activity involves the provision of services rather than the manufacturing of products. Having said that, the characteristics approach can arguably treat services innovation just as well as it treats product innovation. Services have features and characteristics too, and (to a first approximation at least)[6] we can analyse service innovations in the same way as product innovations and indeed, we can analyse combined product/service innovations in the characteristics approach.

A rather special type of product innovation is known in economics as product proliferation. This is the idea that we fill up a product space (or characteristics space) with lots of slightly different versions of the same product. The innovation here is not so much the innovation in any one product, which may be unexceptional, but the strategy of filling the space. We discuss this further in Chapter 5, but briefly, why do companies do it? There are two broad reasons: one unexceptional and the other potentially anti-competitive. The unexceptional reason is that companies find it is profitable to segment markets (as marketers say) and one efficient way to segment markets is to offer a menu of slightly different varieties of the same basic product at different prices. In economics, we would say that this is price discrimination by product differentiation. The potentially anti-competitive reason is that companies seek to fill up the product space so as to make it hard for others to enter. It is rather like the strategy used by some rather selfish train passengers, who place their luggage on the seats around them so as to deter others from trying to sit there. In the same way, a full product space is hard to enter, especially so for the small scale entrant, because any one additional product can only hope to take a tiny share of the market. Some anti-trust cases have suggested that product proliferation can be used in an anti-competitive way.

The final category of innovation that I would list under this heading is *innovative pricing*. It may seem strange to group this with product innovation: innovative pricing does not mean any change to the product or service, but simply means a different way of charging for the service. Nevertheless, innovative pricing is an innovation in terms of what the customer faces in the marketplace and not an innovation in the way something is produced. As such, it is closer to product innovation than process innovation.

What is innovative pricing? It is a new way of charging for a product or service. It is commonly used, for example, by mobile phone operators. Instead of charging a fixed price per minute, the operator may offer a more complicated tariff involving a fixed sum per month, an allotment of free call minutes and free texts, and different prices for on-net and off-net calls.[7] The Oyster cards used on the London Underground are another form of innovative pricing. The regular traveller who has an Oyster card is charged less for the journey than an occasional traveller (a tourist for example) who buys a single ticket for a single journey.

Why do companies use innovative pricing? First, because it offers a subtle way of achieving more effective price discrimination (see Chapter 6). Second, for the same competitive reasons that companies make other innovations: it may allow the innovator to under-cut a rival in a price-sensitive market segment without cutting prices across the board.

Different Types of Process Innovation

We said above that a *pure process* innovation simply changes the way in which a product is made, without changing the product itself. In fact, the literature has identified a variety of different forms of process innovation, including: organisational innovation, supply chain innovation, marketing innovation and what are sometimes called innovations in the 'business model'. For our purposes in this book, however, the economic effects of these are sufficiently similar that they will (at least to begin with) be treated in the same analytical framework.

Other Innovation 'Adjectives'

To complete this section, we define three other innovation 'adjectives' that are in widespread use: *incremental*, *radical* and *architectural*.

Incremental innovation describes the steady stream of improvements to a particular product or process which do not change the character of that product or process in any fundamental way. So, for example, the personal computer has seen a stream of incremental innovations such as: gradual increases in processor speed, gradual increases in memory capacity, and gradual increases in hard disk sizes. But none of these are earth-shattering and, most important of all, these innovations do not undermine the competence of a PC manufacturer.

Radical innovation, by contrast, describes improvements that fundamentally alter the character of a product or process. Moreover, radical innovations are earth-shattering in the sense that they can completely undermine the competence of current market leaders. So for example, the PC was a radical innovation as far as IBM was concerned, and ultimately undermined IBM's dominance of the computer industry (see Chapter 13).

We should not assume that the distinction between incremental and radical innovations is the same as the distinction between small and large innovations. Some incremental innovations may be small but others may be quite large. It is not so much the size of the change that matters, but whether that change is something that the established producers can cope with comfortably (incremental), or whether it undermines the competencies of established producers (radical). Radical innovations may also be large innovations, but as before that is not really the issue. The issue is whether the innovation is disruptive to the established producer. If so, it is radical – even if not very large.

A third 'adjective' is used to describe some innovations: *architectural*. Henderson and Clark (1990) refer to a fundamental change in the way that certain components are pieced together to make a system while the basic components themselves are little changed. Architectural innovations are partly incremental and partly radical. Because the basic components are unchanged, the architectural innovation does not undermine the established firm's knowledge of components. But because the way that components are put together changes, the architectural innovation does undermine the established firm's knowledge of architecture.[8]

PROTECTING INVENTION

We said above that invention was the result of a difficult (and often expensive) process of research and creativity. Most firms will only spend on this process if they are confident that they will be able to achieve some economic return on their investment. That requires two things. First, that there is a reasonable chance the invention will in due course be of commercial value. Second, that the inventor gets some kind of head start in using the invention for economic advantage. If at the end of this difficult and expensive process the inventor finds that his/her invention can be copied easily by rivals, then that will usually undermine the economic return.

To ensure that they can gain the economic benefits from their inventions, firms often take steps to protect their inventions – or their intellectual property, as it is often called. There are two broad ways of doing this. One is to obtain some official mark of recognition that the invention is the property of the inventor and may not be used for commercial advantage by others. The other is to keep the invention strictly secret – at least until the inventor is in a position to turn the invention into an innovation and maybe longer, in the case of a process invention.

The main formal methods of intellectual property protection are patents, registered designs, trademarks and copyright. There are important differences between these (discussed in Chapter 8) but their economic effects are similar – at least to a first approximation. All four of them make the invention public but ensure (through the force of law) that no one else may use or copy the invention. If a rival does violate a patent, trademark, registered design or copyright, then the inventor may sue for civil damages. The effectiveness of these formal methods varies from country to country, however. They offer pretty strong protection in countries such as USA and UK but weaker protection in some other countries.

The patent gives a firm a monopoly right to commercial use of a particular invention embodied in a product or process for a given period (in the USA, this is twenty years from the date of filing the patent).[9] The rationale for this is that in the absence of a patent, some inventions might be copied comparatively freely by many firms other than the originator, and the inventor would not recoup enough to cover the costs of his/her invention. In such a setting the incentive to invent would start to decline, and that would be a bad thing for the long-term prospects of the economy.[10] The aim of the patent is to sustain the incentive to invent. From an economist's point of view, this is more important than the sheer protection of intellectual property. When the originator has recouped his/her costs and a reasonable profit has been made, then, from the economist's perspective, the patent has served its purpose and could (or even *should*) lapse.

The most common informal methods of intellectual property protection are ensuring product/process complexity, wise use of lead-time, confidentiality agreements and other strategies for secrecy. Again, there are important differences between these (see Chapter 8) but the basic idea is to keep the invention secret for as long as possible, or ensure that products and processes are so complex that even if some fragment of information were to be leaked, it would be of no value to a rival without all the other product/process details.

Sometimes companies may take a rather different attitude to their intellectual property. In the open innovation movement (the best known example being the Open Source software movement), companies may share their intellectual property with others even if that means they lose some of the potential economic benefits from their intellectual property. Strictly speaking, open source does not mean an absence of intellectual property rights. The developers of intellectual property retain rights over their property but allow others to use it too. It is a bit like the country landowner who allows people to walk on his/her land so long as they do no damage. This action does not make the land 'common land': the landowner is still definitely the owner. But (s)he does not assert a monopoly over the use of the land for recreation.

The reasons why firms take this open innovation approach are various and can be quite complex. We shall revisit this in Chapter 8.

TWO DIFFERENT THEORIES OF CREATIVITY

While economics has rather little to say about the theory of creativity, and the most important insights come from other disciplines like psychology, I believe it is important for the economics student to have a very basic idea of where creativity comes from. There are many theories, of course, but in terms of their economic characteristics there are two of special relevance to this book. One, derived from the work of psychoanalytical theorist Otto Rank, asserts that exceptional creativity requires the creative mind to develop complete *autonomy*. In this state of autonomy, the creative mind neither knows nor cares what his/her critics may say. Until the creative mind reaches this state of autonomy, the creative person may suffer deep conflict between the desire to fulfil creativity and the painful disapproval of peers. This autonomous theory of creativity is important because it suggests that creative individuals and creative teams may need a degree of isolation.

That appears to be in strong contrast to a second theory, which I shall call the combinatorial theory of creativity. This, which stems from the work of Arthur Koestler and Nobel Laureate Herbert Simon, asserts that exceptional creativity calls for an ability to bring together habitually incompatible ideas and combine them in a way that gives deep new insights. This combinatorial theory is very different from the autonomous theory. Combinatorial creativity calls for an ability to work along the boundaries between conventional disciplines. And since it is hard for any one person to master two disciplines, this sort of creativity may call for researchers and creative people who work well in a team and do *not* act in a deeply autonomous way.

These two theories will be discussed in much greater depth in Chapter 9. In reality, as we shall see in Chapter 9, these two theories are not as directly opposed as it appears above. But they do serve to identify a fundamental dilemma in the theory of creativity. Sometimes creativity calls for people with an ability for (and opportunity for) networking. Sometimes creativity calls for isolation. We shall see in Chapter 13 that different innovation episodes in the personal computer industry can illustrate *both* of these two different aspects. Clusters, as discussed in Chapter 13, can provide an ideal ground for combinatorial creativity. But sometimes radical innovations need to take place outside the cluster, in relative isolation, or otherwise they would never happen.

TWO DIFFERENT VIEWS OF THE CONSUMER

At the right hand side of Table 3.1, we have listed two views of the consumer: the passive consumer and the active consumer. The reader may wonder why it is necessary to discuss the theory of the consumer in a book on innovation. The answer is simple. As we so often find in economics, market outcomes depend on the balance of demand and supply, and that applies to innovation. Much of what follows in this book is about the supply side and how that determines the scope for innovation. But it is essential to have a look at some different perspectives on the consumer because within these we shall find a strong variation in the demand for innovation.

At one extreme, there is a view of the consumer as a creature of habit, who likes to buy the familiar and does not like change. Such a passive consumer will be pretty resistant to any form of product innovation, though may be more interested in those process innovations that allow him/her to buy his/her familiar goods at reduced cost.

At the other extreme is the most active form of consumer, which I shall call the Marshall Consumer – after the great economist Alfred Marshall, who was one of the first to write about this type of consumer. This active Marshall consumer is always seeking variety and change, and is always keen to explore what (s)he can do with new products, even if (s)he has no previous need for them.

In the middle is the conventional consumer of economic theory. This consumer has fixed, pre-determined needs and is only interested in innovation to the extent that process innovations reduce the prices (s)he has to pay, or product innovations to the extent that they provide more of the characteristics that (s)he wants. This consumer has no interest in completely new products which, by definition, (s)he has no need for. And this consumer has no interest in the sort of exploration on which Marshall's consumer thrives.

Depending on the proportions of these different consumer types in the market, there will be greater or lesser scope for companies to sell their innovations. That is why a discussion of the consumer, whether passive or active, is essential in this book. We shall turn to these issues in Chapters 15 and 16.

MEASURING INNOVATION

We finish this chapter with a topic that was not listed in Table 3.1, but which is so important it must be discussed (albeit briefly) in a chapter on basic concepts. How do we measure innovation? The great scientist, Lord Kelvin famously said (1883):[11]

> When you can measure what you are speaking about, and express it in numbers, you know something about it; but when you cannot measure it, when you cannot express it in numbers, your knowledge is of a meagre and unsatisfactory kind: it may be the beginning of knowledge, but you have scarcely, in your thoughts, advanced to the state of *science*.

Kelvin was talking about exact sciences, and some would say economics can never be like physics and chemistry. Nevertheless, it is important for the student to have a basic idea of how we quantify innovation. I would list five basic approaches to measuring innovation that are used in economics.[12]

The first approach is to describe innovations in detail. This is often done in specific case studies or specific industry studies. The advantage of this approach is the depth and detail it offers. The shortcoming is that it is simply too time-consuming for one individual to write a lot of descriptions of this sort. Such descriptions will often contain a lot of engineering data on innovations. An obvious example is the data on computers: processor speed, memory size, hard disk size, etc.

The second is to count innovations. This was done, most famously, in the celebrated SPRU survey of innovation.[13] This survey counted all the important innovations introduced by companies in different sectors of the economy. To do this required a series of sector experts to monitor the trade press and other sources for stories about new innovations. Over time this built up a fascinating picture of the innovation-intensity of different sectors.

The third approach is the questionnaire survey of innovative companies. The best known of these, at present, is the Community Innovation Survey (CIS), carried out in each country within the EU. The British CIS4, which covered the period 2002-2004, had responses from more than 18,000 firms with information on their innovation strategies. Although this does not provide the innovation counts of the SPRU database, it does allow us to paint a pretty detailed picture of innovation intensity in different sectors of the economy. The main shortcoming of such data is that it may be subject to respondent errors and subtle forms of respondent bias.

The fourth approach is to count the number of patents issued. This is relatively easy to do, because Patent Offices (like the US Patent Office, the European Patent Office and the UK Patent Office) have detailed online databases listing all the patents on their books. It is a relatively straightforward task, though quite time-consuming, to create from this data a picture of patent intensity by sector. The main problem with the patent as a measure of *innovation* is that it is *not* a measure of innovation. It is a measure of *invention*, and the two are *not* the same. The patent is awarded to a novel invention, but that invention may or may not ever be commercialised. Indeed, it is generally reckoned that most patents cover inventions of pretty low commercial value. Companies have a wide variety of strategic reasons for patenting other than protecting their most valuable intellectual property – these are discussed in Chapter 8. And there is a further problem with the patent as a measure. While in some sectors (pharmaceuticals and chemicals) the patent is almost always used to protect important inventions, in many other sectors, the patent emerges as one of the least important methods of protecting inventions. We shall look at more evidence on this in Chapter 7. In short, the existence of patents does not necessarily imply innovation and the absence of patents does not necessarily imply the absence of innovation.

The fifth approach to measuring innovation is to use company accounting data on research and development. All large companies in the UK have to declare their expenditures on R&D in their accounts. Smaller companies do not have to but many do, especially in high-technology sectors.[14] As noted above, R&D is – for most companies – really D rather than R, and as a result it is a measure of what is spent on developing innovations out of inventions. There are three problems with R&D as a measure of innovation, however. First, in some small companies there may be no R&D entry in company accounts, but this does not mean that there is no innovation. Second, and related, success in innovation does not just call for expenditure on R&D but calls for other parallel expenditures on design, training, investment and so on. High R&D on its own does not imply high innovation. Third, companies in some sectors have been concerned that a high R&D figure in their accounts may have an adverse effect on the company's stock-market valuation. The reason for this is that R&D is seen as a risky investment. Investors in stock markets will be pleased if the company is seen to do enough R&D to survive in a competitive market but will be anxious if they perceive that the company is spending *too much* on R&D. As accounting conventions give companies some flexibility as to whether expenditures related to innovation are classed as R&D or something else, then companies have some scope to get the R&D figure in line with market expectations. To the extent that this is a problem (and we cannot know how serious a problem it is), R&D may in part reflect what the company wants to signal to the stock market.

In conclusion, we note that the Department for Innovation, Universities and Skills of the British government has (in 2008) commissioned NESTA to produce an innovation index. This will be an ambitious project but, if successful, should provide some further insights into measuring innovation.[15]

NOTES

[1]   This is the definition favoured by those in charge of innovation policy for the British government, see http://www.berr.gov.uk/dius/innovation/

[2]   We find this businessman's definition an especially interesting one in view of some academic work on product innovation as a dimension-increasing activity – e.g. Swann (1990).

[3]   Or, sometimes, the economics of *technical* change.

[4]   Some would say the same of the *iPod*: it used technology that was not really new, but the real innovation was in the design.

[5]   Innovations that add new characteristics are sometimes said to increase the *dimensions* of product space. We shall see in Chapter 5 that there are very specific circumstances in which a dimension-expanding innovation will be important.

[6] Some academic authors (especially those outside economics) make much more of the distinction between products and services. Certainly these do have somewhat different economic properties but for the purposes of the present introductory book we shall not make a lot of the distinction. Much of what we say applies equally to innovation in products and innovation in services.

[7] An on-net call is when you call another mobile phone on the same network as your own mobile. An off-net call is when you call another mobile which is on a different network from your own mobile. Thus if X and Y use Vodaphone while Z uses O_2, then a call from X to Y is on-net but a call from X to Z is off-net. On-net calls, in the UK at least, are typically cheaper than off-net calls.

[8] We could also identify the counterpart to architectural innovation, where the basic architecture is unchanged but the detailed construction of the components does change. This kind of innovation would undermine an established firm's component knowledge but *not* its architectural or system knowledge. However, as far as the author is aware, this counterpart has not yet been given a specific name in the literature.

[9] The basic idea of a patent dates to Roman times. It was used then to provide an incentive for culinary innovation by chefs. If the chef came up with a particularly delicious invention, he would be given an exclusive right to make it for a while (one or two years). Thereafter, other chefs would be allowed to copy. The attractive prospect of a temporary monopoly provided a strong incentive for chefs to be innovative, but the lifetime of the 'patent' was kept relatively short so that no chef could 'rest on his laurels'.

[10] Again this is a classic example of the trade-off between short-run efficiency versus long-run efficiency.

[11] Here quoted from *The Oxford Dictionary of Phrase, Saying and Quotation* (1997, p. 377).

[12] This is not an exhaustive list, by any means.

[13] SPRU is the *Science Policy Research Unit* at the University of Sussex, UK, founded by Christopher Freeman, one of the pioneers of the study of innovation. The SPRU Innovation Survey was discontinued in 1984.

[14] The British government publish an R&D Scoreboard giving data on company expenditures on R&D, at http://www.berr.gov.uk/dius/innovation/randd/

[15] NESTA maintains a website with the latest details of research on this innovation index: http://www.innovationindex.org.uk/

4. Process innovation

A *pure process* innovation simply changes the way in which a product is made, without changing the product itself (except perhaps the price at which it will be sold). In practice, many process innovations are not 'pure' in this way. Often a new and improved process will lead to incidental improvements in the product. Nevertheless, it is helpful to understand the simple economics of the 'pure' process innovation.

PROCESS INNOVATIONS AND COST CONDITIONS

A very simple way of representing the economic effects of a process innovation is to show what it does to production costs, as described by the cost curves. These are the total, average and marginal cost curves of basic first-year economics, and it will be assumed that the student is familiar with these.

Here we shall consider four different types of process innovation and their associated effects on cost curves. These are:

- A reduction in fixed costs, with no change in marginal costs
- A reduction in marginal costs, with no change in fixed costs
- A reduction in marginal costs accompanied by an increase in fixed costs
- A reduction in the marginal cost *of an additional model*.

The first is a basic capital-saving innovation. This could be a reduction in the cost of an essential piece of capital equipment (e.g. a computer). We shall see that such an innovation reduces economies of scale, and allows the small scale company a better chance to enter and survive in the market.

The second is a basic input-saving innovation. This could be a reduction in the use of raw material inputs or an increase in the efficiency with which raw material inputs are turned into final products. Both of these would mean that input costs constitute a smaller share of average cost, and therefore apportioned fixed costs constitute a greater share of average cost. This means, as will become clearer below, that economies of scale are increased.

The third is a typical characteristic of many process innovations where a labour-intensive process is replaced by a capital-intensive process. Fixed costs increase but marginal costs fall. These effects, taken together, provide an important source of increased economies of scale.

The fourth is an innovation that generates economies of scope rather than economies of scale. It describes an innovation that reduces the marginal costs of producing additional variety of output.

Reduced Fixed Cost

The first type of process innovation is the one where the fixed costs of production are reduced. For simplicity we shall focus on the case where marginal costs are unchanged. Figure 4.1 shows how this reduction in fixed costs affects average cost. The upper curve (labelled Average Cost 0) shows the picture before innovation. The lower curve (labelled Average Cost 1) shows the picture after innovation.

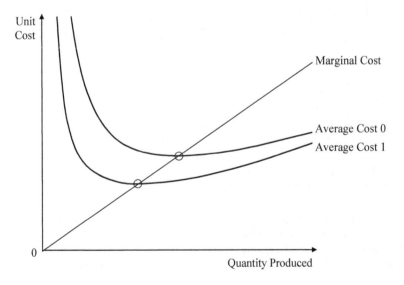

Figure 4.1 Process innovation that reduces fixed costs

We see that the average cost curve is pushed downwards, particularly at lower levels of production. Since we know that marginal cost always cuts the average cost curve from below at minimum average cost, we can say that this process innovation reduces the minimum efficient scale and hence reduces economies of scale.

Reduced Marginal Cost

Now we turn to an innovation that reduces marginal costs while leaving fixed costs unchanged. Figure 4.2 shows how this reduction in marginal costs affects average cost. The upper curve (labelled Average Cost 0) and upper line (labelled Marginal Cost 0) show the picture before innovation. The lower curve (labelled Average Cost 1) and lower line (labelled Marginal Cost 1) show the picture after innovation.

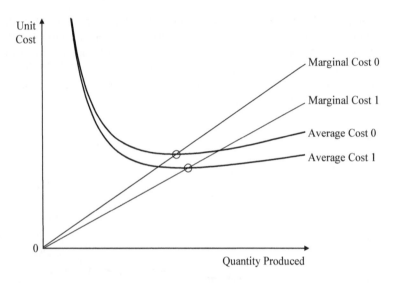

Figure 4.2 Process innovation that reduces marginal costs

We see that the average cost curve is pushed downwards, particularly at higher levels of production. Since we know that marginal cost always cuts the average cost curve from below at minimum average cost, we can say that this process innovation increases the minimum efficient scale and hence increases economies of scale.

Capital-intensive Process: Reduced Marginal Cost, Increased Fixed Cost

One quite common sort of process innovation will reduce marginal costs while increasing fixed costs. Examples of such innovations include the replacement of a labour-intensive production process with a more capital-intensive form of production. Figure 4.3 illustrates this.

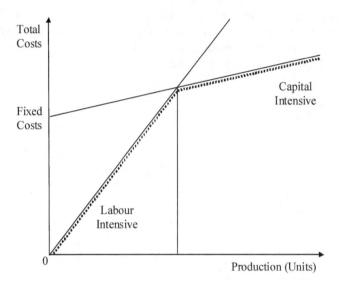

Figure 4.3   Process innovation that increases capital-intensiveness

The labour-intensive process has a low intercept (zero fixed costs) and a steep slope (high marginal cost). The capital-intensive process has a high intercept (high fixed costs) and a shallow slope (low marginal cost). At a low scale of production, the labour-intensive process can produce the required scale of output at lower cost. At a high scale of production, the capital-intensive process can produce the required scale of output at lower cost. The boundary between these two is marked by a vertical line in Figure 4.3. To the left of that, labour-intensive is best; to the right, capital-intensive is best.

The labour-intensive process, as drawn, has constant returns to scale. As the total cost line is a straight line, total costs are proportional to scale of production and hence average cost is a constant. The capital-intensive process, when it is efficient to use it, introduces economies of scale and hence average cost starts to fall.

Flexible Manufacturing: Reduced Marginal Cost, Increased Fixed Cost

A fourth type of process innovation is described by the general term, flexible manufacturing. The flexible manufacturing system increases economies of scope. It does this by reducing the marginal cost of output variability – that is, the marginal cost of allowing one additional brand or model.

Figure 4.4 illustrates this. It is based on the following assumptions. The vertical axis represents the total costs of producing a given number of products (say N). The horizontal axis represents the number of different varieties (or models) amongst those N products. At the left end of the horizontal axis, all the N products are the same. As we move to the right, the variety of products amongst the N increases.

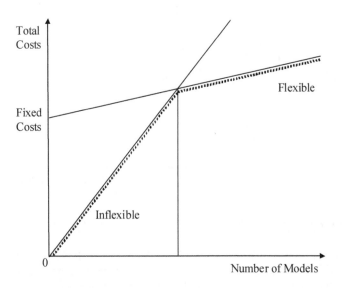

Figure 4.4   Process innovation that increases production flexibility

The inflexible production technology enjoys lower fixed costs but a higher marginal cost per brand or model. The corresponding cost line has a lower intercept but a steeper slope. By contrast, the flexible production technology has a higher fixed cost but enjoys a lower marginal cost per brand or model. The corresponding cost line has a higher intercept but a flatter slope. When little variety is required, the inflexible process is the more efficient. When much variety is required, the flexible process is preferred. As in Figure 4.3, the switchover occurs at the vertical line in the middle of Figure 4.4.

These are perhaps the four simplest forms of process innovation. However, no discussion of basic economic analysis of process innovation would be complete without three further categories. These were very influential in the early literature on innovation which tried to capture all the effects of technological change within a production function. They have become less influential today, because most of those who study innovation do not seriously believe that the effects of innovation can all adequately be captured within a production function.

An innovation is *Hicks-neutral* if the effects can be captured in a production function as follows:

$$Y = \phi(T)F[K, L] \tag{4.1}$$

where Y is output, K is capital, L is labour and T is a variable measuring technological advance. An innovation is *Harrod-neutral* if the technological change is labour-augmenting (that is, it helps labour). In that case, the effects can be captured in a production function as follows:

$$Y = F[K, \lambda(T)L] \tag{4.2}$$

Finally, an innovation is *Solow-neutral* if the technological change is capital-augmenting (that is, it helps capital). In that case, the effects can be captured in a production function as follows:

$$Y = F[\kappa(T)K, L] \tag{4.3}$$

ECONOMIES OF SCALE AND SCOPE

In two of the above examples, innovation leads to an increase in *economies of scale*. In one of the others, innovation leads to *economies of scope*. It is useful for the student to have a basic understanding of these concepts and where these economies of scale and scope come from.

We can define economies of scale and scope roughly as follows. There are economies of scale when average cost declines as the scale of production increases. And there are economies of scope when average cost declines as the scale of production increases, *even if increased production consists of new brands or models.*

More precise definitions would be as follows. Consider the production of a product X. If the cost of producing n units of X is C, and the cost of producing $2n$ units of X is less than $2C$, then there are *economies of scale*. Now consider also the production of Y, which is a similar product to X, but not identical. If the cost of producing n units of X (alone) is C, and the cost of producing n units of Y (alone) is C, but the cost of producing n units of X and n of Y (together) is less than $2C$, then there are *economies of scope*.

In the economics literature, it is common to distinguish between two generic types of scale economy: physical and financial (sometimes called 'pecuniary'). Physical economies of scale, which shall concern us most in this chapter, arise when production processes show increased efficiency when run on a large scale. This can arise because of substantial fixed costs (economies of scale) or *common* fixed costs (economies of scope). Financial (or pecuniary) economies of scale arise when the large firm uses its size to negotiate lower input prices or better inputs, or can use its market power to set higher output prices.

Focusing now on physical scale economies, it is useful to distinguish three categories of scale economy:

1)   Product-specific economies
2)   Plant-specific economies
3)   Multi-plant economies.

The first category is the group of scale economies that are associated with the total volume produced of a particular product. The second category relates to the total output (of perhaps several different products) made at one particular plant. And the third are those economies enjoyed by a company that operates several plants.

To understand the concept of *product-specific* scale economies consider the difference between custom manufacture and production-line manufacture. In custom manufacture, a skilled worker will use a general purpose machine to run off a small batch. The set-up time is minimal, even if the time taken to machine one product is a good deal longer than could be achieved on a production line. In production-line manufacture, a specialised machine will be programmed and used. The set-up time (including programming) would be considerably higher, but when this is done the time taken to produce product thereafter is much smaller. Moreover, the labour input required is less skilled – and hence less costly. Unlike custom manufacture, where the average time taken (and hence cost) per product is essentially independent of the number to be produced, here the *average cost* per product falls as the total batch size increases – essentially because the *fixed cost* of setting up the production line is diluted by a larger total output. This simple example illustrates one of the generic sources of scale economies: a fixed cost divided by an increasing scale of output.

Plant-specific scale economies are most obvious in the context of chemical and metallurgical process industries – such as petroleum, the synthesis of chemicals, iron and steel, and cement. The output of a production unit in such industries is roughly proportional to the *volume* of the processing facilities, while the investment cost of the facilities is nearer to being proportional to the *surface area* of the facilities. From elementary formulae we know that volume is proportional to the cube of radius (r^3) while surface area is proportional to the square of radius (r^2), and hence area (or cost) is proportional to the two-thirds power of volume.[1] In a number of other respects (e.g. fuel consumption) production costs in such industries rise more slowly than in proportion to production volume.

When a *single* plant enjoys economies of scale in producing several different products *above* those that would be enjoyed by a group of different plants producing *one each* of these different products (and in the same volume), then we can say that the single plant enjoys *economies of scope*. In this case it is cheaper for plant 1 to produce x units of X and y units of Y than it would be for plant 2 to produce x of X *and* plant 3 to produce y of Y.

Multi-plant economies of scale apply where the cost of producing x in plant 1 and y in plant 2 are less when these two plants are owned *by the same company* than if the plants are *separately* owned. They arise when the multi-plant enterprise can employ a richer diversity of talent than the single-plant company. The multi-plant enterprise may be able to spread its production, market and financial risks over a larger value of production, and thus reduce its risk and cost of capital. It may also be able to get more out of sales and marketing expenditure.

Learning Curves

The learning curve is a particular sort of scale economy – usually applying at the product-specific level, though it may also apply at the plant-specific level. It is a *dynamic* sort of scale economy in that it arises from a *history of production* rather than the *current scale of production*. The learning curve principle asserts that the unit cost of production for a product declines with the accumulated experience of production of that product. It is usually described as follows: the unit cost of production is measured to fall by x per cent with each doubling of the cumulative volume of production of the product. For many products, x may be of the order of 10-15 per cent, but for semiconductors and aircraft production, x has often been of the order of 30 per cent.

INNOVATION AND SCALE ECONOMIES

In the postwar period, rapid technological change has changed the production methods and costs of a wide range of industries. But has this technological change served to increase economies of scale? One influential tradition of economic analysis (see Chapter 18) has argued that technological change does indeed tend on average to increase economies of scale and hence lead to *increases in industrial concentration*. This is thought to have been particularly important in industries such as steel, aircraft manufacture, chemicals and petroleum. An important distinction needs to be made between process and product innovation. Writers in this *concentrating* tradition argue that process innovation very frequently leads to greater economies of scale, while some product innovations open up niche markets in which smaller firms may be more competitive.

Another tradition (see Chapter 18), however, concludes that while technological change until about 1950 was mostly concentrating, change since then has frequently reduced economies of scale, and has hence been deconcentrating. Examples of this contrary trend include: the substitution of individual electric motors for centralised motors; the appearance of easily fabricated plastics and light metals; the replacement of specialised machines by computer-controlled general purpose machines; and the gradual displacement of water and rail transport by road transport.

This remains one of the open questions in economics. And in addition to the effects of technological change on production economies of scale, it is important to note that rapid technological change can impose substantial managerial costs on organisations. Many writers in the sociology of organisations have argued that smaller (*organic*) companies tend to cope better with unexpected technological change than do large (*mechanistic*) companies. This would provide further support for the thesis that rapid technological change may reduce economies of scale.

VARIETY OF PROCESS INNOVATIONS

The literature on process innovations has developed a much wider vocabulary to describe some of the different generic types of process innovation. Examples include:

- Business Model Innovation
- Marketing Innovation
- Supply Chain Innovation
- Organisational Innovation.

In addition, the field of production management has developed a very rich vocabulary to describe the many different process innovations that have been adopted at different times. These include:

- Fordism
- Taylorism
- Lean Production

- 'Just in Time' Production
- Flexible Manufacturing Systems
- CAD-CAM.

We do not have space in this introductory book to cover all of these. I leave it as an exercise for the advanced student to explore to what extent these particular process innovations can be described using the relatively simple constructs of this chapter.

NOTES

[1]   Area $\alpha\, r^2 = [r^3]^{2/3}\, \alpha$ volume$^{2/3}$

5. Product innovation

A *pure product* innovation creates a new or improved product for sale *without* any change in the production process – except that more inputs (labour, machine time and materials) may be required. In practice, a new product will often require some innovations in the production process, just as a new and improved process often leads to incidental improvements in the product. Nevertheless, the conceptual distinction is an important one.

The framework presented in this chapter is capable of analysing the economic effects of product innovation and service innovation equally. Indeed, within this framework, both these sorts of innovation operate in a similar manner. This is not to imply that they are the same thing and, as noted in Chapter 2, there are more subtle versions of the framework used here which bring out the difference between them. However, that is beyond the scope of this introductory textbook.

CHARACTERISTICS, QUALITY AND PREFERENCES

Economics uses the characteristics approach to analyse questions of product choice and product innovation. Although this approach was first discussed by Gorman and Ironmonger, it is most commonly associated with the work of Lancaster (1971). The characteristics approach treats the product as a collection of features or characteristics. This makes it possible to analyse what happens to markets when improved versions of existing products are introduced. It is very simple on paper, as we shall see. But it is rather harder to apply in practice because it is necessary to count a large number of characteristics to do justice to most real products. While it is easy enough to draw a map of competing products when no more than one or two characteristics are important, that cannot be done (except within a computer) when products are truly multi-dimensional – meaning that they embody a large number of characteristics. Another complication, in practice, is that the dimensions of product space tend to expand over time. This is a phenomenon with important economic implications (as we see briefly below) but it is not easy to handle analytically.

In general, we distinguish between three types of characteristics. The first type is *intrinsic* characteristics – that is, characteristics embodied in the product. These include many dimensions of quality, performance, reliability, features, design and style, and so on. The second type are *perceptual* characteristics – that is, characteristics that are not embodied in the product in a physical sense, but are 'attached to' the product by branding or advertising. The third type is *extrinsic* characteristics, often measuring the quality of the service element provided with the product. This includes delivery, service and support, and a variety of indirect network effects (see Chapter 7).

Within the characteristics framework, we can distinguish product innovations of differing degrees. The simplest is an improvement in one characteristic only. The second is an improvement in several characteristics. The third would be the introduction of one new characteristic. And the fourth would be the introduction of so many new product characteristics that we have, arguably, a completely new product.

The literature on characteristics and product differentiation also recognises one other important distinction, between *vertical* and *horizontal* differentiation. When two products are vertically differentiated, we say that one is unambiguously better than the other. If two personal computers A and B are identical except that B has more memory than A, then we could say that B is (vertically) superior to A. When two products are horizontally differentiated, on the other hand, we can only say that they are different, but we cannot say that one is superior to the other. For example, a red Skoda Fabia and a green Skoda Fabia (identical apart from colour) are horizontally differentiated, meaning that they are different. But we cannot say that one is superior to the other: it depends on one's tastes. Some prefer red to green, some prefer green to red and some others are indifferent.

Indeed, colour gives a good indication of some of the difficulties we can encounter in analysing horizontal product differentiation. Suppose we were to try to define consumer preferences over the traditional visible colour spectrum: violet, indigo, blue, green, yellow, orange, red. Figure 5.1 below gives one possible pattern of preferences.

It is almost impossible to assert any general principles about preferences. Even if I prefer blue to green, we cannot say that in general I prefer colours to the left of the spectrum over colours to the right. I may prefer blue to green, but I may prefer blue to violet and prefer red to orange. Preferences may, as drawn above, be highly non-linear. Moreover, my colour preferences are probably not independent of what sort of object we are talking about. I may prefer a red car to a black car, but that doesn't mean I would prefer a red suit to a black suit.

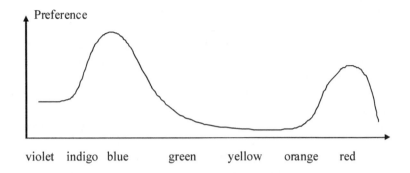

Figure 5.1   Possible preferences over the visible colour spectrum

For these reasons, economic analysis of horizontal product differentiation can be quite difficult. In what follows, therefore, we shall focus only on vertical differentiation, which is rather easier to analyse.[1]

In most cases of vertical differentiation, we can say that most people prefer better to less good (if the price is the same), though a few may not care about quality and are hence indifferent between better and less good. People differ however in the intensity with which they prefer better to less good. Some don't care much and are not prepared to pay much for superior quality. Some care a lot and are prepared to pay a lot. We shall see how this simple observation can be made operational through what is called a *willingness-to-pay* (WTP) curve.

PRODUCT MAPS, CHOICE AND TERRITORY MAPS

For graphical simplicity, we shall analyse the case where different varieties of a product can be compared in terms of one characteristic alone (and price).[2] Figure 5.2 below illustrates the case of three products. Thus, product A represents the *cheap and cheerful* product, product C the premium product, and product B the mid-market product. In addition we have drawn a typical willingness-to-pay (WTP) curve. This describes how much the consumer is prepared to pay for increases in quality. Apart from the likelihood that they are upward sloping, it is difficult to generalise about the shape of these curves. It is however common to find that below a certain quality, willingness-to-pay drops off quite sharply (as shown here) – indicating that the consumer insists on a certain minimum level of quality, and below that the product is of little value.

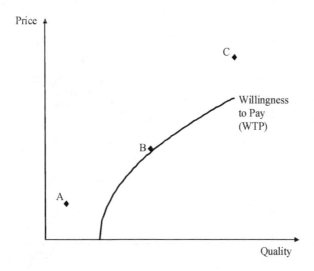

Figure 5.2   Willingness-to-pay and product choice

It is also common to find that for many consumers, willingness-to-pay tends to level off at high quality – indicating that beyond a certain point, typical consumers become satiated, and place little value on further improvements in quality. There are of course exceptions to both of these generalisations – especially the second, as the existence of *premium* brands and products will testify. Nevertheless, this general shape of willingness-to-pay curve is common, even if the curves for different consumers may be located in different parts of the diagram.

In this case, product A does not appear attractive to this particular consumer, since (s)he is not willing to pay the asking price. While cheap, it does not appear to be good value for money. At the other end, product C, though of high quality, is too expensive in the eyes of this consumer. But product B looks the best value for money to this consumer, since the asking price is about equal to what the consumer would be willing to pay for it. In a variety of circumstances, we would expect a well-informed and rational consumer to choose product B in preference to A or C in this setting. That is, the consumer would choose the product offering best value for money, which is priced at or below the consumer's willingness-to-pay.

Now, although WTP is often curved as shown, it is useful shorthand in what follows to make the assumption that WTP lines are *linear*. This means they look as shown in Figure 5.3 overleaf.

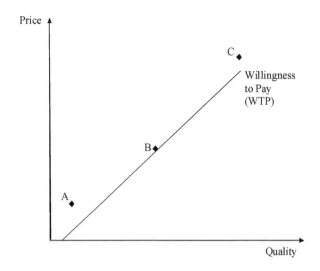

Figure 5.3   Willingness-to-pay and product choice (linear WTP)

In this case, again, the consumer with the WTP line as shown would normally choose product B. The other two products are just too expensive for the quality they offer. This assumption of linearity allows us to draw what are sometimes called, 'product territory maps'. Figure 5.4, below, shows the territory map corresponding to Figure 5.3.

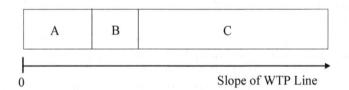

Figure 5.4   Product territory map corresponding to Figure 5.3

The territory map shows, for different slopes of the WTP line, which product will be chosen. So, if the WTP line is flat (slope = 0), then the consumer is not willing to pay for increased quality, and therefore buys just the cheapest product (A). Now consider what happens as the slope of the WTP increases. Eventually we reach a point where the consumer is indifferent between A at its lower price and B at its intermediate price. As the slope of the WTP increases beyond that, B is preferred to A. Then eventually we reach a slope where the consumer is indifferent between B at its intermediate price and C at its high price (while preferring either to A). Then as the slope increases beyond that, C is the clear first choice. The territory map captures this in a simple way. It shows, for each slope of the WTP line, the consumer choice from A, B and C.

The product territory map as shown is unexceptional. It just captures the idea that product A will be bought by low-end consumers, product B by mid-market consumers and product C by premium consumers. That is also pretty obvious from Figure 5.3. However, the product territory map comes into its own when we use it to show the effects of product innovation on consumer choice. We turn to that next.

PRODUCT AND PROCESS INNOVATION COMPARED

In Chapter 3, we made much of the distinction between product and process innovations. The analytical framework described above can illustrate their different effects on the product market. To see this, consider Figure 5.5. The initial pattern follows Figure 5.3, but two possible innovations are added.

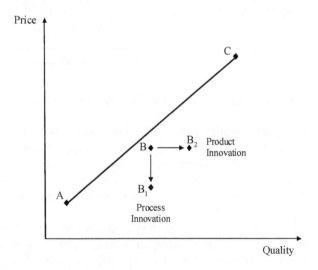

Figure 5.5   Product and process innovation

If the producer of B implements a cost-saving process innovation in the production of B, then it would be possible for that producer to relocate B to a reduced price (B_1). We shall describe this strategic move as using a cost-reducing process innovation to cut price. Alternatively, if the producer of B can achieve a product innovation with no addition to costs, then it is possible to relocate B to a higher quality (B_2). The move to B_1 brings B closer (both in terms of the diagram and in economic terms) to product A. Intuitively, we would expect this price reduction to mean that B cuts significantly into the market share of product A. On the other hand, the move to B_2 brings B closer to product C. Intuitively, we would expect this quality increase to mean that B cuts significantly into the market share of product C. The product territory map confirms these intuitions.

Figure 5.6 shows the product territory maps before any innovation (the middle row), after the product innovation (top row) and after the process innovation (bottom row).

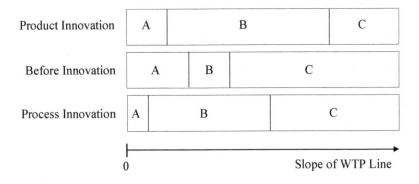

Figure 5.6   Product territory maps before and after innovations

Compared to the pre-innovation picture, both innovations allow product B to capture a larger market share. But they achieve this increased market share in different ways. The process innovation takes market share from A *and* C: the territory for B expands more or less equally in both directions. The product innovation, by contrast, mostly takes share from product C and much less so from A.

As drawn, the difference between the top line and bottom line in Figure 5.6 may not seem great. But in more complex settings with more competing products and more dimensions of quality, the difference in effect of product and (cost-reducing) process innovations can be very substantial.

PRODUCT PROLIFERATION

Product proliferation is a special type of product innovation. It is the practice of proliferating a wide variety of slightly differentiated products across the entire characteristics space. Figure 5.7 illustrates this.

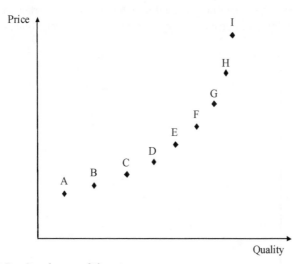

Figure 5.7   Product proliferation

We shall discuss the rationale for this in a moment, but first, let us see what it does to the product territory maps. Figure 5.8 shows the territory map corresponding to Figure 5.7.

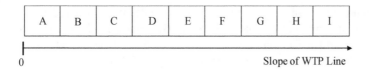

Figure 5.8   Product territory map in case of product proliferation

This proliferation of products breaks up the consumer space into small territories or segments. There are two reasons why a firm might wish to do that. The first reason is that this is quite an effective method of market segmentation. The marketing technique of market segmentation aims to break the consumer base into different groups and to set different prices in each in order to increase profitability. In economic terms, we can show that this sort of product proliferation allows the company to achieve an ever more efficient form of second-degree price discrimination (for more on this see Chapter 6).

The second reason for product proliferation is rather different. Sometimes companies find it worthwhile to proliferate a much larger number of models – not so much to segment markets as to deter entry by others. Proliferation of products can deter small scale entry because a single product entrant in a congested marketplace can only expect to achieve a small market share, and may not cover the fixed costs of entry. To see this, consider Figures 5.9 and 5.10.

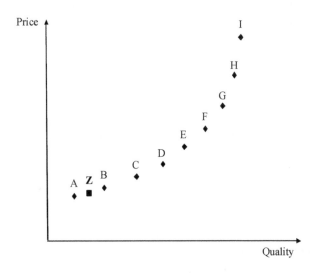

Figure 5.9   Product proliferation deters small scale entry

In Figure 5.9, a small scale entrant tries to enter this congested market by offering only product Z (in between A and B). Although this product is priced so that it will make some sales, its share of the market is pretty small, as shown in the corresponding territory map (Figure 5.10).

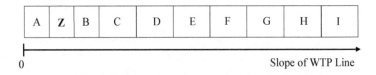

Figure 5.10   Small scale entrant takes only a small niche in territory map

It is rather like the physical analogy of a person trying to board a very congested train. There is only a small 'breathing space' and it is not comfortable. Seeing this, the would-be entrant may simply abandon plans to enter the market. It is in this way that proliferation may deter entry.

However we should make two observations about the limitations of product proliferation as an entry deterrent. First, while product proliferation by an incumbent may deter small scale entry, it does not necessarily deter large scale entry. The large scale entrant can also proliferate different models, and while any one of these may not win enough market share to cover fixed costs of entry, the collection of models may earn enough revenue to cover these fixed costs. In effect, the entrant uses product proliferation as an *entry strategy*. Second, as we shall see in the next section, product proliferation does not necessarily deter the entrant who adds extra characteristics (or increases the dimensions of characteristics space).

One of the striking examples of the use of product proliferation as an entry deterrent is in the ready-to-eat breakfast cereal market – see below. One of the striking examples of proliferation as a successful entry strategy was when Japanese motor-cycle manufacturers entered the US market (Kotler et al., 1986).

PRODUCT PROLIFERATION IN BREAKFAST CEREALS

The case of breakfast cereals is often cited as an example of how product proliferation may have acted as a deterrent to entry. It is a well-known case because of the complaint pursued by the US Federal Trade Commission against the four largest US manufacturers of ready-to-eat breakfast cereals, and because it is the subject of two famous articles by two of the best-known industrial economists: Schmalensee (1978) and Scherer (1979). The FTC complaint suggested that proliferating brands and differentiating similar products, along with intensive advertising, had resulted in high barriers to entry. The complaint was eventually dropped, but it helped to develop our understanding of when product proliferation is beneficial to consumers, and when it is not.

Breakfast can mean many different things to different people. While the focus of this case is on ready-to-eat breakfast cereals, it is recognised that there are many other breakfast foods. Where do we draw the boundaries of the market for the present case? Those who have studied the ready-to-eat cereals market have argued that if the choice between ready-to-eat breakfast cereals and other foods is made on grounds of taste or habit (rather than price), then it is reasonable to treat ready-to-eat cereals as a sub-market in their own right. This is to assume that the cross-price elasticity of demand for ready-to-eat cereals with respect to the price of other breakfast foods (and vice versa) is low. If these cross-price elasticities are low, then the existence of alternatives does little to compensate for the loss in consumer welfare that results from an increase in prices in the ready-to-eat market.

Scherer (1979) notes that the SIC industry defined as 'Cereal Preparations' is one of the most concentrated US manufacturing industries. The three firm concentration ratio has been 80 per cent or above for many decades. Profits after taxes as a percentage of assets for the leading firms averaged out at 19.8 per cent compared to 8.7 per cent for all manufacturing corporations. All of this evidence suggests that this is a highly concentrated and profitable industry. Market structure has also been pretty stable over time. The early entrants to the market were Quaker Oats, Post (later a part of General Foods), Kellogg, and from 1928, General Mills. Since the 1950s, Kellogg's market share has been about 45 per cent, with General Mills at about 20 per cent and General Foods at about 15 per cent. Few markets in the US (or elsewhere) have shown such stability in market shares.

For many observers, the introduction of numerous new product variants was the most striking aspect of this case. Scherer (1979) states that in 1957, the six largest manufacturers had between them a total of 38 brands in national distribution. By 1970, the figure was 67. Shepherd (1990) states that by 1973 the figure was 80, and that many more have been introduced since then.

Many of the new brands failed to achieve viable market shares, and indeed of the 51 brands launched between 1958 and 1970, only five achieved market shares of 2 per cent or more. Most of these new products filled only small niches in the market, while many of the older brands showed considerable staying power. Only two pre-1958 brands were withdrawn from the national market during the 1960s. Moreover, the 29 brands that led the market in 1960 (with 83 per cent of industry sales) continued to account for 74 per cent of the market in 1970 – despite the proliferation of new competing brands.

If these new brands were relatively unsuccessful, why did the major firms continue to introduce them? One argument is that the large manufacturers enjoy economies of scope, and that the break-even market share required for an additional cereal variety is well below the break-even market share required for a one-product company. From this perspective, the proliferation of products looks reasonable enough, and to the extent that proliferation increases the variety available, it can be a good thing from the consumer's point of view.

A second argument could be that in a crude comparison between old and new brands, the average old brand is bound to look more successful because the unsuccessful old brands have already been withdrawn from the scene, while the average new brand is an average of successful and unsuccessful. In a competitive oligopoly, each player has to continue to innovate to stay competitive, and the appropriate form of innovation in this context is the new cereal variety. Such innovations will contain a mix of failures and successes, but as the old maxim says, 'if you're not losing money on some lines, you're not trying'.

A third argument would accept the points about economies of scope and the risks inherent in innovation, but would go further. The firm's incentive to proliferate brands is not simply to expand its sales. It is also to deter entry. If competing products can be thought of as points in a product space, then the more products there are, the more congested the space becomes. Just as a congested geographical space can be unattractive to enter, so is a congested product space – essentially because in a congested space, firms can only expect to sell to a limited niche, and so the prospective sales from entry are small. If incumbent firms proliferate a sufficient number of slightly different cereals, then they can ensure that the only remaining niches are too small for the entrant to achieve the necessary break-even level of sales.

NEW CHARACTERISTICS/DIMENSIONS

Product differentiation by introducing new product characteristics becomes increasingly attractive as a competitive strategy when a characteristic space becomes congested – as for example in the case of product proliferation. Demand becomes less price-sensitive as the product is differentiated. As a result, we tend to observe that producers tend to increase the dimensions of product space (add new characteristics) when it gets congested.

To see this, consider the discussion around Figures 5.8 and 5.9 above. When the space is congested, a single product entry cannot expect to take much market share, and cross-price elasticities between incumbents' products and the entrant's product will be high. In that context it is attractive to try to make more competitive space.

Figure 5.11 shows this. This is derived from Figures 5.7 and 5.9, but dropping the price axis and introducing a new characteristic on the vertical axis. All incumbent products A-I are clustered along the horizontal axis. But if the entrant can introduce an additional product characteristic (Y) as shown and, assuming that this characteristic is valued by at least some consumers, then the entrant can expect to gain a larger market share and reduce the cross-price elasticities between demand for his/her product and other products.

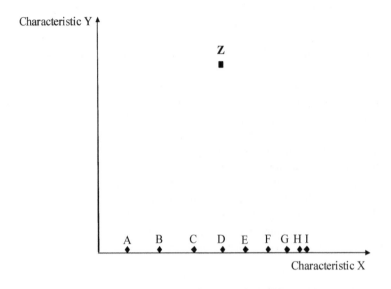

Figure 5.11   Entrant introduces new product characteristic

The two-dimensional product territory map corresponding to Figure 5.11 is a little complex to derive, but it would show a substantially larger market niche for Z in this case than the territory map in Figure 5.10.

PRODUCT CHOICE BY THE PRODUCER

So far, we have used the characteristics approach to analyse consumer choice. We can also use it to analyse the producer's choice of what product specification to offer. To carry out this analysis we need to superimpose a cost function (describing cost as a function of quality) on our willingness-to-pay (WTP) curve. (Here we revert to the curve of Figure 5.2.) It should be possible using production engineering information (or otherwise) to estimate an approximate relationship between cost of production and product quality.

Figure 5.12 below shows what we might find. It is hard to generalise about the shape of this cost curve, other than to say that it is upward sloping. Nevertheless, the curvature shown in the diagram is quite common, suggesting that at high levels of quality, the cost of further improvements in quality may be much greater than at moderate levels of quality.

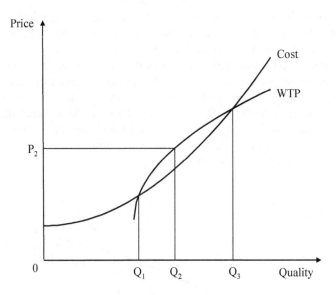

Figure 5.12   Producer choice of product quality

Any quality between Q_1 and Q_3 would be a possible choice for the producer in that the consumer's willingness-to-pay for this variety would equal or exceed the cost of production. If the producer were concerned with maximising his/her profit margin in this context, then product Q_2 would be chosen, and sold at price P_2. This is the quality at which the willingness-to-pay exceeds cost by the largest possible amount.

FLEXIBLE MANUFACTURING AND PRODUCT VARIETY

We can use our framework to answer a second question about the producer's decision. How many different models should the producer choose to sell, and how does flexible manufacturing change the answer?

A full analysis of this question is beyond the scope of this chapter, but the basic ideas can be described here. Essentially it depends on the balance between two factors. If consumer tastes are very diverse, then it may well be that the producer needs to make a wide variety of brands in order to maximise sales, and the revenues from sales. On the other hand, there is a fixed cost in producing any additional brand so there is also an incentive to limit the number of brands produced to restrain costs. The balance between these factors will determine how wide the portfolio of products should be. We can summarise the story in one further diagram (Figure 5.13).

This shows how total costs (including brand launch and production costs) increase as the number of brands increases. The shape of the cost line reflects a substantial additional cost for each additional brand.

The graph also shows how the expected sales revenue increases as the number of brands increases. The shape of the sales revenue curve reflects the fact (as we argued above) that two brands can exploit the profit potential of different market segments better than one, and ten can do so better than two, and so on. Nevertheless, beyond a certain point there are diminishing returns to increasing the number of brands, as the existing group can cover the diverse market segments adequately well.

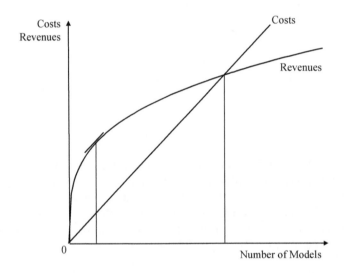

Figure 5.13   How many models to produce?

In the graph as shown, the producer has quite a wide range of profitable choices, where revenues cover or exceed costs, but the profit-maximising number of models is relatively small (shown by the left-hand vertical line).

Now we can ask what happens if we replace the cost line shown above by the cost line for a flexible manufacturing process, as illustrated in Chapter 3. We saw in Figure 4.4 that the cost line for such a flexible manufacturing process has a higher intercept but a lower slope. It is a more costly capital-intensive process to set up, but once it is installed, the company can proliferate the number of models produced at relatively low marginal cost per model.

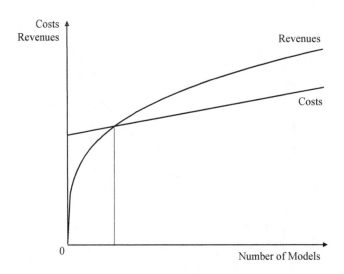

Figure 5.14   How many models to produce? (flexible manufacturing process)

In this case, there are a minimum number of models before the company breaks even, but beyond that, the company can proliferate an almost unlimited variety. Even at the right-hand side of this diagram we have not reached the most profitable number of models, still less the break-even number of models. Comparing Figures 5.13 and 5.14, we can see how this flexible manufacturing process allows the producer to market a far greater degree of product variety.

MEASURING PRODUCT COMPETITIVENESS

We conclude this chapter by showing how the above analytical framework can be used to measure the competitiveness of individual products. To see this, we take the basic structure of Figure 5.3 and add some more products (D-H) as shown in Figure 5.15. We also draw an *envelope* around the lower boundary of the various products. The rationale for doing that will become clear in a moment.

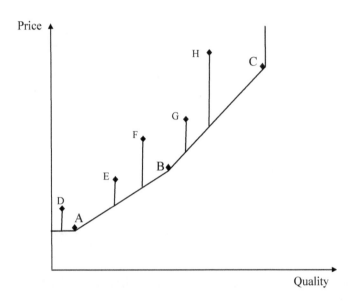

Figure 5.15   Measuring product competitiveness

Now, suppose that WTP is linear (as in Figure 5.3). Then we know from the territory map of Figure 5.4 that all consumers will buy one of the following: A, B or C. The consumer with a WTP line given by AB will be indifferent between A and B. If the WTP line is flatter, (s)he will buy A. If steeper, (s)he will prefer B. Likewise, the consumer with a WTP line given by BC will be indifferent between B and C. If the WTP line is flatter, (s)he will prefer B. If steeper, (s)he will buy C. Whatever the slope of the WTP line, nobody will buy any of the other products D-H.

We can now ask ourselves: what would have to happen to the price of product E (say) for someone to want to buy it? The answer is that the price would have to fall until it reaches the envelope. This means the point at which the vertical line below E crosses the envelope. At that price, a consumer with a WTP line given by AB would be indifferent between A, B and E. This is the highest price at which E can expect to sell. In the same way, we can compute the maximum prices at which all the other products (D, F-H) can expect to sell as the price at which that product would just lie on the envelope.

This practice of computing an envelope or efficient frontier is known as *Data Envelopment Analysis* (DEA). It was originally used to compute productive efficiency but it is equally applicable to measuring product competitiveness. If a product lies on the envelope or frontier, then we can say it is competitive. There are at least some consumers who may buy it. In Figure 5.15, only three products are on the frontier (A, B and C). If a product lies inside (above) the envelope or frontier, then we can say it is uncompetitive. It cannot expect to find a buyer because any consumer, whatever the slope of his/her WTP line, will prefer one (or more) of the products on the frontier. The extent to which these products inside the frontier (D-H) lie above the frontier (i.e. the vertical line from these products to the frontier) is a measure of how over-priced the product is. That is a natural measure of the extent to which it is uncompetitive.

Amongst the products inside the frontier, it is worth distinguishing them into two groups. Products D, F and H are obviously uncompetitive in the sense that there is another product on the market which is better and cheaper. A is better and cheaper than D; B is better and cheaper than F; and C is better and cheaper than H. Economists say that A *dominates* D; B *dominates* F; and C *dominates* H.

However this observation does not apply to E or G. There are no products that are simultaneously better and cheaper than either of these. Why then are they uncompetitive? They are uncompetitive because any consumer (with a linear WTP) would prefer A and/or B to E, and any consumer (with a linear WTP) would prefer B and/or C to G. As a point of fine detail, if we were to consider WTP *curves* such as in Figure 5.2, then it is possible that some consumers would have WTP with enough curvature so that they prefer E to A or B, or prefer G to B or C.

The above gives a simple intuitive (and not very mathematical) explanation of how DEA is used to measure product competitiveness. For a more detailed discussion of the mathematics of DEA, and how it is applied to the measurement of product competitiveness, see Swann and Taghavi (1992).

An older approach to measuring product competitiveness is called the 'Hedonic' method. In this method, a parametric (often linear) regression line is estimated, with price a function of quality. The regression residuals for each product can be used as measures of the competitiveness (if negative) or uncompetitiveness (if positive) of each product. This method has several shortcomings, however – see Swann and Taghavi (1992).

NOTES

[1] When we apply horizontal differentiation to geographical space, then it is often easier to define preferences.

[2] The earliest approaches to characteristics analysis use a two-characteristic diagram just like the two good model of basic consumer theory. The snag with that is that either: (a) we have to assume that all products have the same price (unlikely): or (b) we have to assume that product characteristics can be continuously scaled up and down. So, for example, this is akin

to assuming that we can break up a Ferrari car to give us several Trabant cars (unlikely) or that we can combine several Trabant cars to make a Ferrari (very unlikely).

6. Innovative pricing

Innovative pricing can be defined as the activity of creating new pricing schemes or tariff structures. Pure innovative pricing is where such new pricing schemes are introduced with no (perceptible) change to the product or service. The traditional term in economics for this phenomenon is *price discrimination*: charging different prices to different customers for the same product or service. But more recently, it has inherited the more exciting label, *innovative pricing*, to reflect the fact that some companies put a lot of innovative effort into devising such pricing schemes.

Why do companies engage in innovative pricing? They do so because it is profitable, or because innovative pricing offers a way to extract sufficient revenue from a market to cover costs. In this chapter we shall show how innovative pricing can be interpreted as a form of price discrimination. Innovative pricing schemes which are then maintained in a stable form for some time can be seen as a form of discrimination (or segmentation) by customer characteristics or by product/service quality. Innovative pricing schemes which are constantly subject to change can be described as a form of 'noisy' price discrimination.

Loosely speaking, price discrimination occurs where a producer sells the same product or service to different buyers at different prices. What is the rationale for price discrimination? Stated simply, the rationale is that some buyers are willing to pay more for a particular product or service than others. It is therefore profitable to design a pricing scheme whereby those prepared to pay a lot are charged a lot, while those prepared to pay only just above cost are charged just above cost. So, for example, if consumer A has an elastic demand curve and consumer B has an inelastic demand curve, it is profitable to charge B a higher price than A. In essence, this rationale is the same as the marketing rationale for market segmentation – and indeed, you could argue that all schemes of price discrimination depend on market segmentation.

PROBLEMS WITH PRICE DISCRIMINATION

There are three potential problems with price discrimination. The first is the legal one: in general there is a presumption in law that price discrimination acts against the consumer interest. The second is a philosophical problem with the definition. The third is the issue of whether it is ethical.

The legal issues are complex, and well beyond the scope of this chapter. As Shepherd (1990, p. 488) observes in the US context, 'Systematic discrimination by dominant firms was often held to be illegal ... Sporadic price discrimination by firms with small shares was rarely challenged.' At the same time, the welfare economics of price discrimination (which does, presumably, underpin some legal thinking on this) is complex too. One view is as follows: 'Generally, discriminatory prices *will* be required for an optimal allocation of resources in real life situations' (quoted from Phlips, 1983, p.1, my emphasis – though note that Phlips himself does not necessarily subscribe to this view).

Turning to the philosophical problem, some observers would argue that if two different sales are done at different prices, then they must represent *different* (even if only *slightly* different) products or services. From this perspective, a safer definition is that price discrimination occurs when a producer sells a similar product or service to different buyers at different prices. In some of the examples we see below, it is the fact that the product or service is slightly differentiated that enables the producer to charge different prices.

This is turn opens up yet more difficulties on the legal side. If different products are sold at different prices, is that necessarily discriminatory? Perhaps the differences in price reflect differences in cost? In anti-trust discussions, attention sometimes shifts from discriminatory price differences to discriminatory differences in profit margins (i.e. price *minus* cost): in that case, *discrimination* occurs when a producer makes a higher profit margin from one sale than another. According to this definition, indeed, it could be discriminatory to set equal prices! For example, it could be argued that *uniform* postal tariffs (independent of distance) discriminate against users of local mail and in favour of users of long distance mail – because it must be cheaper to send a letter one mile than one thousand miles. Many, however, would see this as an example of socially desirable discrimination.

Now we turn to the third question: is price discrimination *ethical*? That is a very hard question to answer because it depends on which ethical perspective you take, and what is the alternative to innovative pricing.

Let us start by contrasting two popular perspectives on ethics. In Kantian ethics, the fundamental principle is that we must fulfil what Kant called the categorical imperative. This means that we should:

1)   only accept principles which we would accept as a universal law
2)   respect humanity as an end in itself
3)   act as if we are both buyers and sellers at the same time.

The utilitarian perspective on ethics is rather different: whether an action is right or wrong is decided by its consequences, and *not* by the inherent character of the action. Actions which achieve a beneficial outcome are good; those which do not are not.

How do innovative pricing and price discrimination look from these two ethical perspectives? From the Kantian perspective, price discrimination does not always look very attractive. Are we really content to see price discrimination as a universal law? Noisy price discrimination seems unattractive because it exploits the ignorance of some customers. And any price discrimination based on the fact that those *in extremis* have no choice but to pay higher prices seems doubly unattractive. Is it really humane to charge an air passenger a higher fare because (s)he has to travel urgently to be with a loved one who is seriously ill? And while sellers may use price discrimination in their businesses, can they honestly say they never object when they face price discrimination as customers?

From a utilitarian perspective, however, the picture can look a bit different. Consider the case where the fixed cost of running a flight is high and the only way an airline can break even is by price discrimination. In this case, we face a harsh choice: without price discrimination there is no flight, and some customers definitely lose out; with price discrimination, some customers pay more than others, but at least the flight is available. From this utilitarian perspective, therefore, price discrimination and innovative pricing may be a necessary evil – without which socially and economically valuable services cannot break even and will not be offered.

CONDITIONS FOR PRICE DISCRIMINATION

Two conditions need to be satisfied if price discrimination is to be used profitably. First, there must be different willingness-to-pay (or price elasticities) in different markets. If not, then price discrimination will not be profitable. Second, the sorting devices used in price discrimination must be successful at sorting consumers into the required different groups. Price discrimination will break down if all those who have been targeted to pay a high price manage to get served in the low price market.

DEGREES OF PRICE DISCRIMINATION

It is traditional to recognise three degrees of price discrimination.

First Degree (Perfect)

Here, each buyer is charged the maximum they are willing to pay. This means in effect that no buyer makes any profit (or consumer surplus) on the purchase, but instead the seller makes all the profit.

This is a rather idealised form of price discrimination, for two reasons. First, because it will always be difficult to establish a particular buyer's maximum willingness-to-pay. It certainly isn't in a buyer's interest to reveal it to the seller! Second, because first-degree price discrimination requires in effect a different price for each person – and it is difficult to see how a seller could develop a subtle enough pricing scheme to cater for this.

Some would argue that *haggling* between an experienced seller and a naive buyer can approximate to first-degree price discrimination. The seller works out just how much the buyer would be prepared to pay, and gets that price! It is conceivable also to envisage first-degree price discrimination in *Dutch auctions* (where the price called starts high and is reduced until the buyer makes a bid). The determined buyer might bid as soon as his/her reservation price is reached.

In anything less than first-degree price discrimination, it is recognised that the producer cannot expect to set a different price for each buyer. At best, the producer can set a discrete number of prices in different settings, or perhaps define a pricing rule that results in a variety of different prices for different customers. But the producer cannot hope to extract all the profit. Some customers at least end up paying less than the maximum they would be willing to pay.

Second Degree

The simplest definition of second-degree price discrimination is where a producer sets a number of different price tags for the product or service, and each customer ends up paying the highest of those price tags consistent with their maximum willingness-to-pay. So for example, if prices are set at 10, 8 and 6, then the customer willing to pay up to 11 actually pays 10, the customer willing to pay up to 9 actually pays 8, and the customer willing to pay 7 ends up paying 6. In this way, price tags are closely matched to the consumers who are willing to pay that amount, but not much more than that.

The *key points* about second-degree price discrimination are these. First, each buyer ends up with the highest price tag that they would accept. Second, even those who are not willing to pay much will still get a chance to buy so long as they are prepared to pay the lowest price tag.

Third Degree

Here, the producer sells a product in several different markets (for example, different regions) and charges a different price in each. Usually, the high prices are set in markets where demand is inelastic while the low prices are set where demand is elastic. But unlike the case of second-degree price discrimination, there is no guarantee in this case that each customer ends up paying the highest price tag consistent with their maximum willingness-to-pay. In terms of the above example, some of the customers willing to pay 11 end up paying 10, but some will only pay 8 or 6. Conversely, some of the customers willing to pay 7 will not get a chance to buy at 6, because they are unfortunate enough to live in a high price region.

The main difference between second- and third-degree price discrimination is that the two key points listed above do apply to second degree but don't apply to third degree.

Between Second and Third Degree

Some observers consider that there are price discrimination schemes that lie in between second and third degree, in that they satisfy one of the *key points* listed above, but not the other. Consider a scheme where a producer sells a standard brand at price 10 and a premium brand at price 15. Setting aside whether this is discriminatory in a legal sense, it can be interpreted as a means of price discrimination if premium buyers willing to pay a high price for a premium product select to pay 15, while economy buyers (not willing to pay a high price) select to pay 10. So long as customers select brands and prices in this way, then this is second-degree price discrimination.

Suppose instead that some premium buyers, who are perfectly willing to pay 15 for the premium brand, nevertheless choose that they would be better off buying the economy brand at 10. Then this is not really second-degree discrimination because it cannot be said that each customer is paying the highest price tag consistent with his/her maximum willingness-to-pay. So is this an example of third-degree price discrimination? No, because no consumers need be locked out of the market because they are stuck in a high price market rather than the low price market. Anyone willing to pay 10 and above can buy.

In practice we tend to classify such examples as second degree, even though they do in a sense belong to an intermediate category. The point is that if such pricing schemes are well designed, the amount of *trading down* will be limited.

SOME GENERIC EXAMPLES OF PRICE DISCRIMINATION

What follows is not an exhaustive list – see Shepherd (1990) or Phlips (1983) for further details. But these are some of the main generic methods of price discrimination.

Two-Part Tariffs

In the two-part tariff, the buyer pays a fixed cost and a usage-related cost. So, for example, the domestic telephone subscriber pays a fixed rental charge, and then a charge for each call made. This means in effect that heavy users pay a smaller average cost per call since the fixed cost is diluted over more calls. This can be interpreted as price discrimination, if the heavy user is price sensitive and the light user is price insensitive: the price elastic user is charged a lower cost per call than the price inelastic user. Figure 6.1 illustrates how these two-part tariffs work for mobile phones. It shows total cost as a function of call minutes, for three different tariffs offered by one operator. As the graph shows, total cost can vary significantly according to which tariff is chosen. The dotted line shows the 'best deal'. (In technical economic language, it is the lower *envelope* of these three tariffs.)

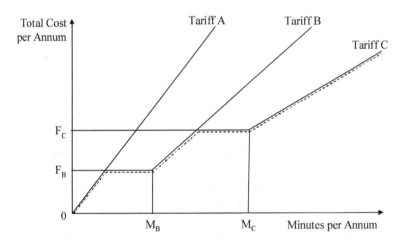

Figure 6.1   Mobile phone tariffs

Tariff A makes no fixed fee but has a relatively high charge per minute. Tariff B has a fixed fee of F_B and M_B free minutes. Above that, call charges are lower than in Tariff A. Tariff C has the highest fixed fee (F_C) and in return offers M_C free minutes. But it also has the lowest marginal cost per minute for call minutes in excess of M_C. The occasional user would be best to choose Tariff A, while the heavy user is best to choose Tariff C. Tariff B would be the best choice for the intermediate user, between these two extremes.

Other examples of two-part tariffs are the entry fee to an amusement park, club membership charges, and so on. Quantity discounts offer a similar pricing structure to the two-part tariff. Notwithstanding the point raised above, two-part tariffs are generally treated as second-degree price discrimination, as indeed is any other scheme where average price is determined by the intensity of customer use of (or demand for) the service.

Pricing According to Consumer Characteristics

Students often pay a lower price than the standard price. The same applies to pensioners. The marketing logic of this is that students as a group tend to have a lower maximum willingness-to-pay (or ability to pay) than those in full-time employment, and student demand is more price elastic than the norm. Such price discriminatory schemes are widespread (e.g. travel cards, museum entrance fees, journal and club subscriptions). They should probably be classed as third-degree price discrimination, because there are some (a few?) rich students who are quite capable of paying the regular price, and conversely there are some in full-time employment who would only buy at the student price – but they are debarred from that market.

Pricing Over Time

Peak-time travel is more expensive than off-peak travel. Off-peak telephone calls are typically much cheaper than peak-time telephone calls. Night-time electricity is cheaper than daytime electricity. Holidays in August tend to be more expensive than holidays in February. In some high technology areas, a new product may be introduced first at a high price, but if you are prepared to wait, the price will fall later on. All of these are examples of price discrimination over time: the willingness-to-pay for peak-time travel, high-season holidays, and early delivery of a new product is high, and such pricing schemes exploit this. This form of discrimination lies perhaps between second and third degree, but is usually classified as second degree. Another form of intertemporal price discrimination is *penetration pricing*: the product is initially sold at a very low mark-up on costs (or even at a loss) with the aim of building up market share. Then a higher price is charged when the product is established.

NOISY PRICING

These sorts of pricing schemes are perhaps the most subtle of all. It is suggested that some multi-store retailers may set different prices in different stores to sort out the searchers from the rest. The argument is that the busy customer, whose time is scarce and valuable and does not have the time to search, will pay the first acceptable price (s)he comes across, while the customer with time to spare will do more market research and seek out a lower price. It is also suggested that some producers develop very complex pricing schemes so that those with the time to work them out will seek out a better deal than those who are too busy and confused. Some have suggested that this second strategy may apply in parts of the travel business. Both are examples of price discrimination by noisy or complex pricing. This should probably be classed as second-degree price discrimination.

Financial services offer many examples of noisy pricing. In general, whether we are talking about borrowing or lending, it is generally true that new accounts offer more favourable rates of interest than old accounts. Why? The idea here is that customers tend to be slow to switch from one account to another, even when the latter offers a better deal. Economists use the term 'switching costs' to capture this (Klemperer, 1987). Consumers with time to spare, or who are particularly sensitive to price will switch when it is favourable to do so, while the rest will not. Some financial institutions exploit this by 'churning'. This is the practice of introducing new accounts with better terms and gradually making the old accounts less attractive. In this way, financial institutions can price discriminate by offering better terms to new account holders (who have just switched) and poorer terms to existing customers (with inertia).

Tariffs for mobile phone usage are also subject to noisy pricing. Tariffs change regularly. The consumer may pay considerable attention to these at the time of purchase, in order to work out the best deal for his/her needs. But relatively few customers continually monitor these tariffs to work out whether their existing service still offers the best deal for them – and if not, switch to another. Figure 6.1 above showed clearly that the value for money offered by different tariffs depends heavily on whether the user picks the 'right' tariff for his/her needs. Mobile phone operators can, by this continuing process of tariff innovation, expect to charge higher prices to those with inertia and lower prices to those who are more sensitive to price.

Some have argued that the growth of e-commerce and online shopping will eradicate noisy pricing. After all, noisy pricing depends on a certain degree of ignorance on the part of the consumer. It is a hassle to search around for the best deals, and that is part of the reason why customers don't switch. But the evidence on this is mixed. In a study of prices and price dispersion amongst online bookstores, Clay et al. (2001) found that over their sample period there was no evidence of a reduction in average prices or of price dispersion. This suggests that even if online shopping makes price comparisons easier, that is still not working through to less 'noise' in pricing. Part of the reason may be that online sellers sometimes make it time-consuming to dig out details of their pricing structures. This is especially true with online sites selling train tickets. That in turn makes it time-consuming to perform price comparisons. However, these search costs are declining with the emergence of web-sites that act as search engines over a number of online bookstores.

RECENT EXAMPLES OF INNOVATIVE PRICING

While the Internet may make it easier for the consumer to work around noisy prices, it also offers the supplier great opportunities to create new innovative pricing schemes.

One example, which has attracted a lot of interest, is Priceline (www.priceline.com). This online site sells airline tickets, hotel rooms, rental cars, home finance, and so on. There is nothing exceptional about that – except its pricing model. Instead of the supplier quoting prices for a particular product or service, the consumer is asked to state his/her own price. So, for example, the consumer requests a particular journey by air and states a price which (s)he is willing to pay for that journey. Priceline require some flexibility on the part of the consumer. In particular, the consumer must commit to buy at that price without complete knowledge of the deal: (s)he does not know exactly which airline will carry him/her or the precise timing. But in return for that flexibility, Priceline claim that they can allow consumers to save up to 40 per cent on brand-name products and services. Varian (2000) shows how this is just another form of price discrimination. Those who are price sensitive but product flexible can signal this flexibility by using Priceline. Those for whom travel is time critical but are not price sensitive are better to buy their tickets from conventional outlets.

Is this scheme all that innovative? Some would dispute that. Indeed, some commentators were amazed that this pricing scheme was awarded a US patent for this 'reverse auction' model. Moreover, Priceline has encountered its fair share of controversy, since consumers may not be aware of just how much flexibility is expected of them. This pricing model is at odds with one of the fundamentals of the market: that you have a full sight of the product or service you are buying before you commit to buy.

Some online suppliers are believed to operate what is sometimes called dynamic pricing. That is to say, the price is customer-specific and will depend on the customer's history of purchases, amongst other things. In principle, with a long enough history of transactions, the online supplier might be able to use dynamic pricing to achieve something like first-degree price discrimination. However, it is highly controversial, and a leading example of how price discriminatory schemes may generate bad-will amongst customers.

Yoffie and Cusumano (1999) describe how Netscape used innovative pricing methods to become the market-leading Internet browser by 1995. Its innovative pricing model was described as 'free, but not free'. The official price of Navigator 10 was $39, but for educational and non-profit use it was free, and it could be used for a free trial period of 90 days. Netscape probably knew full well that some users would pay up after the trial period, but many would not. But the aim of 'free, but not free' was to build up a large network of users as soon as possible, and (with luck) set the market standard. When established as a standard, Netscape could hope to *sell* the browser (and no longer offer it free). Meanwhile, Netscape would make its money by selling services on their web servers. This 'innovative pricing' mechanism is a mix of different methods of price discrimination: pricing according to consumer characteristics, penetration pricing and a sort of two-part tariff (where services on web servers are sold at a high price while the browser itself is sold at a subsidised price).

WHAT CONSTRAINS PRICE DISCRIMINATION?

Although innovative pricing and price discrimination can be profitable, some suppliers may be reluctant to use them. There are a number of constraints on price discrimination as a strategy.

Regulation

As noted before, price discrimination may be illegal. If price differentials between different products do not appear to relate to cost differentials, the anti-trust authorities may take the view that this is discrimination against the group of consumers who suffer from the higher cost-to-price mark-up.

Competition

Certain types of price discrimination are only viable if the producer has a degree of market power. Competition in general erodes the potential to make monopoly profits, and this includes those monopoly profits made through price discrimination. Competitors will find it attractive to supply the market in which a would-be monopolist is charging a large cost-to-price mark-up.

Arbitrage

As noted above, price discrimination breaks down if consumers targeted to pay a high price manage to buy at the lower price. One reason, of course, is that the consumers who buy at a low price could in principle resell in the high price market – and that clearly undermines price discrimination. Price discriminatory schemes which charge a corporate subscription to companies but a much lower rate to private individuals would break down if private individuals from a company bought the benefits of subscription for the company as a whole, but only paid the low price. Subscription contracts are often written to preclude such outcomes.

Commitment and Reputation

Intertemporal price discrimination – that is charging a high price at launch, but reducing the price later – can be a good mechanism to sort out customers with high willingness-to-pay from those with low willingness-to-pay. It exploits the impatience of consumers, and the competitive advantage for corporate customers in being able to secure early delivery of a new product. But this game will not work indefinitely. If you gain a reputation for intertemporal price discrimination of this sort, then consumers will get wise to it and may delay purchase in anticipation of future price cuts. In such a setting, some producers may decide to make a firm public commitment that they will not cut prices – so there is no point in consumers waiting for the price cut that will never come. But such commitments only work if you keep to them.

CASE STUDY OF INNOVATIVE PRICING: AIR TICKETS

If you were to ask the person sitting next to you on a flight, 'how much did you pay for your return fare?' the chances are that (s)he would not have paid the same price as you. Now, admittedly, even an economist might hesitate to ask this conversation-stopping question! Nevertheless, the question is a fair one: why are there so many different fares for what, at first sight, seems like the same journey?

This phenomenon is not, of course, confined to air fares. We observe the same with many goods and services: In summer 2008, the return train fare from Nottingham to London could be as low as £16 and as high as £198; the price of a package holiday varies according to the time of year and whether you are booking in advance or taking a last minute bargain; the cost of a mobile phone call depends on the specific tariff chosen; some online evendors charge lower prices for their goods at weekends than during the week; and so on. I'm sure the reader could add many other examples, but in what follows I focus on the example of air fares.

How can an airline charge different passengers different fares for what is (in essence) the same product or service? In this context, the words 'in essence' are very important. The fact that two passengers are sitting side by side on a particular flight does *not* imply that they were sold exactly the same thing. Perhaps one traveller was sold an expensive and flexible return that allows him/her to travel on any flight, while the other was sold a cheap but restricted ticket that allows no flexibility. The two passengers were not sold exactly the same thing even if they end up sitting together.

In the same way, airlines can charge different fares according to: (a) the time of sale and the time of travel, (b) where the ticket was bought, especially whether the tickets are sold online or through a travel agent; (c) whether the customer is a student, a leisure traveller or a business traveller; (d) the class of accommodation; (e) whether the ticket sale is bundled with other travel tickets; and many other factors.

All these innovative pricing schemes can be categorised into two broad groups: *systematic* price discrimination and *noisy* price discrimination. In the first category, price differences are well known and well understood but some customers still opt to pay the higher price. In the second category, price differences show an element of randomness and some customers do not know where to find the best bargains. The second category offers another dimension to price discrimination because it distinguishes between those customers with the time to seek out the best price and those busy customers who don't have time to search but will buy so long as the price is reasonable.

Why do companies want to do this? The simplest answer is that price discrimination happens because companies find it profitable. To see this, consider the very basic demand curve in Figure 6.2. This describes how market demand for seats on a particular flight would vary as the airline alters its fares. If the airline charges a high fare (around £200), then demand will be very low, but if the airline charges a very low fare (about £1) then demand will be high.

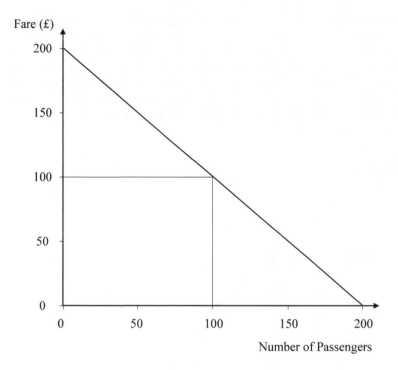

Figure 6.2   Demand curve for air travel

We can also interpret the demand curve in a different way. It tells us how many (and which) customers are prepared to pay how much. Those customers at the left-hand side of the diagram are prepared to pay a lot (up to £200) while those at the right-hand side are only prepared to pay a little.

Now to simplify things, make two assumptions. First, suppose that it costs the airline nothing to transport one additional passenger in an empty seat. This may be a slight exaggeration but not much. Second, suppose that the capacity of the aeroplane is 200 people (plus crew).

If the airline decides to set just one fare, then the most profitable price will be approximately £100, as shown on the diagram, and the revenue raised will be £10,000. But there are two defects in this pricing strategy. First, at this fare the plane may only be half full (100 passengers). There are some potential customers who would be prepared to pay something to travel and it costs nothing to offer them a seat, but the plane leaves without them. Second, there are some passengers who would be prepared to pay higher fares. For these two reasons the airline is not achieving as much revenue as it could.

Suppose, instead, that the airline sets a range of fares, as described above. And suppose it can – by skilful innovative pricing – ensure that those customers willing to pay more end up paying more. Then the airline can extract a good deal more revenue from the market. Figure 6.3 shows how this works in the case of three fares. Passengers 1-50 are charged £150, passengers 51-100 are charged £100 and passengers 101-150 are charged £50. This way, the airline can raise £15,000 from ticket sales – 50 per cent more than when it sets a single fare.

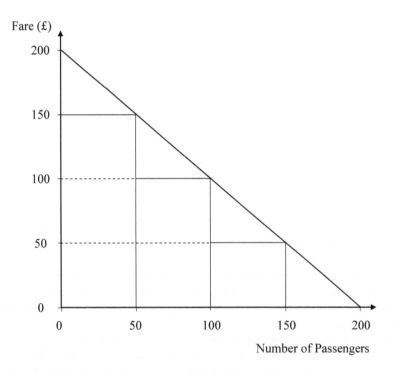

Figure 6.3   Price discrimination: three fares

As a general proposition, the more distinct fares the airline can set the more revenue can be raised. Following similar calculations to those above, Figure 6.4 below summarises how revenue increases with the number of distinct fares.

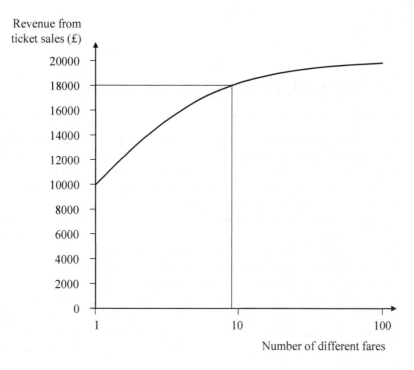

Figure 6.4 Revenue achieved by innovative pricing

In the limit, if the airline can charge each passenger exactly the maximum (s)he would be prepared to pay then it can extract the maximum possible revenue. The real world is never as simple as our simple diagram, but if it were, this maximum revenue is about £20,000 – double what could be raised by setting one fare only. That is why this innovative pricing strategy is so profitable.

However, it is not just desire for profit that leads to such a strategy. In any business where there are large set-up costs (or *fixed* costs) this strategy may be necessary *just to break even*. In terms of our simple example, suppose that the fixed cost of operating the flight was £18,000. The airline has to meet this cost regardless of how many passengers it carries. The airline with just one fare can only raise £10,000, but that isn't enough to cover its fixed costs. The airline charging three prices can raise £15,000, but that isn't enough either. From Figure 6.4, we can see that it requires nearly 10 distinct fares to raise £18,000 and break even, and 10 or more to make a profit. So innovative pricing becomes not just a way of making a large profit but may be essential if the company is merely to break even.

7. Network effects and standards

In traditional economic theory of demand, we generally assume that the value a consumer obtains from consuming a product is independent of whether others also consume the same product, or indeed *how many* others consume it. This may be an appropriate assumption in many settings, but not in all. For example, the value of a 3G mobile phone (i.e. a phone with video) to me depends on whether my friends also have 3G mobile phones and indeed *how many* of them have a 3G phone. If none of my friends have 3G, then it is of little value to me that I can send video because none of them can receive. Nor can I receive 3G video from them. If one friend has 3G, then at least I can make some use of the 3G functionality, but not much. If most of my friends have 3G, then I can indeed make extensive use of 3G functionality. In short, the value of my having a 3G handset depends on how many of my friends also have it. In traditional demand theory, by contrast, the fact that others consume the same product or service neither adds to nor subtracts from its value to me.

This is an example of what we mean by a *network effect* (Rohlfs, 1974).[1] When network effects are important in consumption, we say that the value of consumption is not only a function of the intrinsic characteristics of the product itself but depends also on the number of others who consume the same product or service. In the example of 3G mobile, network effects are *positive*: the more of my friends that have 3G, the greater the value to me. Indeed, most of the literature on network effects makes the assumption that these are *positive*.

We can however find some similarities between the concept of demand for distinction and the idea of a *negative* network externality. If a Premier League footballer seeks distinction by buying a Lamborghini sports car, then the value of that car as a mark of distinction requires that not too many others also own Lamborghini cars. As a first approximation, we could say that if too many people own such cars then a negative network externality would apply: value as a source of distinction *declines* the more people own that type of car. However, in the rest of this chapter, we shall be concerned mainly with positive network effects. We shall discuss the demand for distinction in Chapter 15.

The literature makes a very important distinction between *direct* and *indirect* network effects. When we refer to direct network effects, we mean that the user benefits *directly* from the existence of a large network of other users. So for example, the network effects related to 3G mobile handsets are direct. The benefit I obtain from my 3G handset depends on the amount of 3G communication I can do with it and that in turn depends directly on how many friends have a 3G handset. These *direct* network effects often arise because the user benefits by 'communication' with others that use the same product or service.

When we refer to indirect network effects, on the other hand, we mean that the consumer does not benefit *directly* from the other users of the system. Rather, the fact that there exists a large network of others using the same product or service will convey some *indirect benefit* to our consumer. This could be because the existence of a large network of users means it is more likely that there will also be a large range of *supporting products and services*, and these are of value to the user.

Some further examples will help to clarify the distinction. The following are examples of *direct* network effects:

- the value of a telephone network depends on who else is connected
- the value of an email service depends on who else is connected
- the value of an instant messaging service depends on who else is connected.[2]

In each case, the network effect derives directly from the other users. By contrast, the following are examples of *indirect* network effects:

- the value of an advanced DVD system depends on availability of software (i.e. films)
- the value of a particular computer system depends on availability of compatible software
- the convenience of a particular make of car depends on availability of parts and competent garages (and that depends on the total number who drive that make of car).

Here, the network effect does not derive directly from the other users. I may not care whether others use the same DVD as me; I may not care whether others use the same computer as me; and I may not care if others drive the same model of car as me.[3] However, in each case I benefit indirectly from the fact that there are many other users of the same product. I benefit because the fact that many others use the same DVD mean that there is a large market for software (films). I benefit from the fact many others use the same type of computer because that means there is a large market for computer software. And I benefit from the fact that many others drive the same sort of car, because that means there is a large network of garages that can repair my car, and it probably also means that spare parts (and servicing costs) for my car are cheaper than for rarer types of car. The benefit is indirect, but none the less important for that.

One further type of network effect has been recognised in the more recent literature: the *tariff-related* or *tariff-mediated* network effect.[4] Probably the most familiar example of this occurs in the context of mobile telephones. In the UK, in particular, and also some other countries, it is common to find that mobile phone operators charge a higher tariff for 'on-net' calls than for 'off-net' calls. This means that mobile operator X charges his/her customers a lower per-minute price for calls to other customers of X than for calls to customers of a rival mobile operator (Y). This means that mobile phone networks are subject to tariff-mediated network effects. If I am a customer of X, then it is cheaper for me if as many as possible of my friends are also customers of X, because then I can enjoy these lower rates. This is not quite the same as the *ordinary* direct effect described above. Technologically speaking, I can make a call to any of my friends with their own mobile phone *regardless* of which operator they use. But the network effect shows up in the form of reduced call charges if I use the same operator as do my friends. For that reason, we call this a tariff-mediated network effect.

Are these tariff-mediated network effects *direct* or *indirect*? In general, they are *direct*. The tariff-mediated network effects can be measured as the size of the savings I make because my calls are mainly 'on net' (i.e. to friends using the same operator). The size of the benefit is *directly* related to the number of my friends who use the same operator as I do.

In the next section, we shall examine three widely used rules of thumb that describe how network effects vary according to the size of the network. But before finishing this section, we should introduce one further concept which, though conceptually quite distinct from a network effect, is often especially important in situations where network effects are also important.

SWITCHING COSTS

In the traditional economic theory of demand, we generally assume that the consumer continually selects the best combination of price and quality. If competition or innovation means that a new product offers better value for money than any of the existing products, then the customer will switch without difficulty to the new. In short, we assume that there are no significant *switching costs*.

That is a reasonable assumption in some contexts. For example, suppose that I am in the habit of buying a particular brand of caster sugar from the supermarket, because it is generally the best value for money. Suppose then a new brand becomes available, which is clearly better value for money. While I may have some residual brand loyalty, there is no large switching cost in selecting the new brand in place of the old.

In many contexts, however, the assumption of low switching costs is unrealistic. There are two main reasons for this.

First, the act of switching may involve me in more work than keeping the same supplier. This is relevant to the decision to switch from one electricity supplier to another. Even if we know that our existing supplier does not offer the best value for money, and we have no brand loyalty to the existing supplier whatever, we may be disinclined to go to the bother of switching because it requires several hours of searching on the Internet and a sequence of letters, emails and phone calls with old supplier and new. These are genuine costs and may outweigh the benefits of switching – though sometimes we may overestimate the size of these switching costs.

Second, I may be reluctant to switch when I am familiar with a particular system, or if I have made large investments in it. So, for example, if I am familiar with a particular software package, I may be reluctant to switch to a new package – even if it is better and cheaper. I have built up a pool of software-specific expertise and it is costly to have to do that all over again with a new package. Or, to take another example, I may be reluctant to switch my bank account from one bank to another because I have (as I see it) built up a stock of goodwill as a reliable customer who does not default and so the bank offers me a better overdraft arrangement than I could expect with a new bank which knows nothing of my financial reliability.

When switching costs are important,[5] customers will not continually select the product or service offering the best value for money. This does not mean, at the other extreme, that consumers exhibit total inertia, or a complete reluctance to switch supplier. But it means that such switches are occasional rather than continuous. We shall see below, in our discussion of the 'standards race' that the *combination* of network effects and switching costs can make it much more likely that we observe a form of technological *lock-in* whereby we seem to be stuck using inferior technologies when better alternative exist. Such lock-in is less common if we have either network effects or switching costs on their own.

THREE 'LAWS' OF THE ECONOMICS OF NETWORKS

In the last section we said that network effects refer to the case where the value a user obtains from a particular product or service is an increasing function of the number of others using the same product or service. It is natural to ask whether we can make any more precise statements about the functional form of that relationship. The truthful, if somewhat inconvenient, answer is that we cannot make any broad generalisations about the form of this function. Much of the theoretical literature on network effects assumes that the value I obtain is proportional to the number of other users. That is a convenient assumption, but it will only be true in rather special circumstances (Swann, 2002a).

However, even if we cannot make any broad generalisations, it is useful to summarise three 'rules of thumb' that have emerged from the literature and which describe the relationship between the value derived from a network and the size of that network. These are known, respectively, as Sarnoff's Law, Metcalfe's Law and Reed's Law. It has to be said that although all of these are useful concepts, none of them really has the status of a 'law'. It would be better to describe these as 'rules of thumb', but for economy of exposition we shall use the original expressions in what follows.

Sarnoff's Law states that the *aggregate* value (V) of a network (to all those in the network) is proportional to the number of members in the network (N):

$$V = cN \qquad (7.1)$$

This means that the average value of network membership to any *individual* member (u) is a constant:

$$u = \frac{V}{N} = c \qquad (7.2)$$

This 'law' was proposed to describe the value of some broadcast networks. The total value that an audience obtains from watching the broadcast is proportional to the number of viewers. (We could replace the word 'viewing' by 'viewing or listening' throughout this paragraph, but for economy of exposition stick to just 'viewing'.) This seems like a reasonable assumption in the context of what we might call 'private' viewing – that is, where the value that one person obtains from watching a programme is unrelated to how many others watch it. This might apply to watching the weather forecast, for example. But the assumption would not be appropriate for 'social' viewing, where the value from watching a programme depends on the fact that others are also watching. This might apply to 'soap operas' and certain types of 'reality TV' where the programme is a topic of discussion with friends and colleagues, and hence the value of watching is not independent of how many others choose to watch.

A second law, which is perhaps the best known of the three, is Metcalfe's Law. This asserts that the *aggregate* value (V) of a network (to all those in the network) is proportional to the square of the number of members in the network (N):

$$V = cN^2 \tag{7.3}$$

This means that the average value of network membership to any *individual* member (u) is proportional to the number of network members:

$$u = \frac{V}{N} = cN \tag{7.4}$$

In other words, the value to an individual (u) of a telephone connection depends on the number of (relevant) people (s)he can call. This law is implicit in much of the theoretical literature on network effects (as described above). This 'law' applies to telephone networks, in particular, or other communication networks (fax, email). The rationale is that the value of such a network is proportional to the number of different phone calls between different pairs of callers that it can support. We can show that this is approximately[6] proportional to N^2.

Metcalfe's Law may exaggerate because it assumes that an individual caller is equally likely to call any other. In practice, caller networks are not like that. There are a group of friends and associates that I am likely to call, and then everyone else who I am not likely to call. As my friends and associates join the network then I gain value from that but as others outside that group join the network then I do not gain from their joining the network. In practice, this often means that the first members of a network add the most value, while later entrants add less value. For that reason, individual value is not proportional to *n*, but grows less than proportionately and eventually reaches a ceiling. More generally, the user will pay more attention to the *composition* of the network of other users and not just the *number* of users *per se*.

The third law, and the least well known, is Reed's Law. This asserts that the aggregate value (V) of a network (to all those in the network) is an *exponential function* of the number of members in the network (N):

$$V = c2^N \tag{7.5}$$

What is the rationale for Reed's Law? It is based on the assumption that the total value of a network is proportional to the total number of *distinct groups* or teams of size 2 and above that can be formed within that network. For large N, the answer is approximately[7] proportional to 2^N. Reed's Law may be especially relevant to the formation of creative teams where the required competencies are uncommon and the ability of team members to work with each other cannot be taken for granted. The key here is to form a team with the required competencies and the ability to work together. But if Metcalfe's Law is an exaggeration, then Reed's Law is even more of an exaggeration. Reed's Law assumes that each user attaches equal value to every potential group (s)he could belong to. In practice, it is most unlikely that the network user would attach equal value to each and every possible group. Nevertheless, Reed's Law captures an important idea in creative networks, and may also offer an explanation of why it is relatively easy to form creative teams that work well in large clusters – an issue that we shall discuss in Chapter 10.

IMPLICATIONS OF NETWORK EFFECTS

The existence of network effects can have a variety of implications for economic behaviour. We can group these effects into three categories:

- Implications for consumers of products with network effects
- Implications for sellers of products with network effects
- Implications for companies that wish to exploit the fact that they are part of a network.

We discuss each of these in turn.

Implications for Consumers

When there are network effects in the consumption or use of a particular product or service, then consumer behaviour may change in some important ways. The first implication is that the consumer's choice problem becomes more complex. The consumer no longer makes his/her choice decisions only with reference to the price and intrinsic quality of available products. The consumer will also want to take account of the size (and possibly composition) of the network of others who use the same product or service.

Figure 7.1 illustrates this. Suppose that the consumer faces the choice between two software products with the same price. One of these (A) is of relatively low quality but a large number of other consumers already own this product. The other (B) is of higher quality, but only a small number of other consumers own this product. If network effects did not exist, then since the prices of A and B are the same, the sensible consumer should prefer B. But when network effects are important, then the choice is not so straightforward. The consumer will need to weigh up the relative importance of quality and network effects.

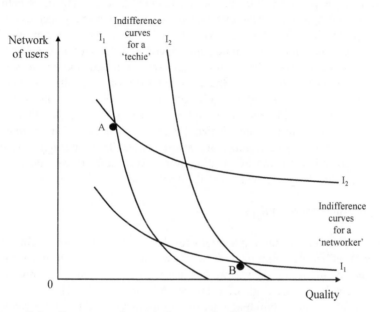

Figure 7.1   Consumer choice with network effects

Figure 7.1 shows the indifference curves for two different types of consumer. The 'techie' is primarily interested in the intrinsic quality, or technological sophistication of the software package, and is not so concerned if only a few others use the product. As a result, the 'techie' has rather steep indifference curves, as shown. By contrast, the 'networker' is less concerned about intrinsic quality, or technological sophistication, but is more concerned to ensure that (s)he buys the same product as is used by friends and colleagues. As a result, the 'networker' has rather flat indifference curves, as shown. In Figure 7.1 we have drawn two indifference curves for each consumer type. The lines lying further from the origin represent a higher level of utility (I_2). The lines lying nearer to the origin represent a lower level of utility (I_1). It should be clear then that the 'techie' prefers product B to A, while the 'networker' prefers product A to B.

The second implication is that when network effects influence choice, buyers may prefer an inferior technology to a superior technology because the former has a large network of users while the latter doesn't. This is especially relevant when a substantial number of consumers are 'networkers'. Moreover, as we shall see below, this observation lies at the heart of why superior new technologies may not displace inferior but established technologies.

The third implication is that a group of consumers for whom network effects are important will try to coordinate their choices.[8] This is sometimes called a 'consumption economy of scale'. We encounter many such examples in everyday life. One familiar example is that members of a family often coordinate their choice of mobile phone operator. This saves money when on-net calls are cheaper than off-net calls (as described above). A second example is that employees in the same company often use the same word-processing software. This makes joint authorship easier because the document files can be passed from one author to the next to make their own contributions. A third example is that different communities often *standardise* on one system to take advantage of network effects. So, for example, business users very often use the PC + Windows + Microsoft Office environment, while graphic designers tend to use the Apple Mac.

Implications for Suppliers

Next we can turn to the implications for the suppliers of products for which network effects are important. If, as shown in Figure 7.1, consumer choice depends on network effects as well as intrinsic quality, then the supplier must try to build up a network of users as soon as possible. The firm that is successful in quickly building a network of users will find that their system is more attractive to subsequent users. As these users join the network that makes the product even more attractive to the next wave of consumers, and so on. In short, there is an important element of positive feedback here – or 'success breeds success'.

In the absence of network effects, different products may co-exist in different segments of the market. But when network effects are important, it is unlikely that many products can survive together. The existence of this positive feedback means that competition with network effects often tends to produce one winner. For this reason, the literature describes the challenge as a 'standards race', to capture the idea that what is at stake is a race to dominate the market.

Later in this chapter, we shall discuss the standards race in more detail and some of the strategies that suppliers use to try to win the race.

Implications for 'Network Firms'

Finally, we can turn to the implications of network effects for what we shall call the 'network firm'. By a 'network firm' we don't necessarily mean a firm that produces products for consumers who value network effects. Rather, we mean a firm that makes it an explicit part of strategy to exploit being part of a network. Network firms are firms that specialise in a very narrow part of the vertical chain, and outsource most other activities. We shall see in Chapter 13 that such network firms are common in strong industrial clusters. In that context, the network firm can use some of the many firms in the cluster to carry out many parts of the value chain, while keeping to itself those activities in which it has a particular skill or expertise. Clearly, this strategy works best when the network firm can draw on a large network.

STANDARDS AND DOMINANT DESIGNS

In the last section we said that one of the implications of network effects for the consumer is that there are incentives for those who live and/or work together to standardise on a common product or service. And we saw that when network effects are important in the demand for a particular product, the supplier of that product faces a 'standards race' to try to achieve market success for his/her product. It is generally recognised that the idea of network effects and the idea of a standard are quite closely related, and this section will define some of the different uses of the term 'standard'.

The most important distinction is between 'formal' and 'informal' standards. Sometimes formal standards are called 'institutional' standards or 'de jure'[9] standards, while informal standards are called 'market' standards or 'de facto' standards (or even 'dominant designs', so avoiding the use of the term 'standard').

Formal standards are public *documents* written by one of the standards institutions (such as ISO, CEN, BSI or DIN) or consortia established with the purpose of writing a specific standard for a specific technology. As these documents are public and open, any company is, in principle at least, at liberty to produce a product or service that adheres to the standard. These formal standards are defined by a committee within a standards institution and emerge as a result of a process of consensus and/or compromise. It can be time-consuming to reach agreement on a formal standard, but at the end we have an open document, and that openness can be important for innovation. In addition, standards professionals believe that this institutional standard produces a result of higher quality than would emerge from competition in the market.

By contrast, informal standards are not public documents and are generally not open. Most usually, these are proprietary designs owned by one or more companies, and their claim to be a 'standard' derives from the fact ('de facto') of their market success rather than institutional endorsement. These 'dominant designs' have emerged successfully from a standards race and have emerged as a standard through use. Allowing such standards to be defined by market competition has several disadvantages. It can be a very costly process, especially for the losers, but also for consumers. This market process of reaching a standard may increase the risk of lock-in to an inferior outcome. Moreover, the end result is a proprietary standard rather than an open standard and that introduces an undesirable element of monopoly. But set against these disadvantages, the market process is generally quite quick at producing a 'winner'.

In the discussion of standards races that follows in the next section, we refer to informal or market-defined standards because these are the result of a standards race. But formal standards are also important for innovation as we shall see at the end of this chapter.

In addition to the formal/informal distinction, the literature on standards tends to distinguish four different types or functions of standards. Any one standard may combine more than one of these functions, but it is useful to make a distinction between them because their economic effects are subtly different.

The first is the compatibility standard or inter-connection standard. The role of the compatibility standard is to ensure that we can connect items A and B, and that they work with each other, or that a piece of software will run on a particular piece of hardware. When we discuss 'standards races', in the next section, it will be this type of standard that we are concerned with. Compatibility standards exist to ensure that the user of a particular product who values network effects can indeed enjoy those network effects. These compatibility standards can be formal or informal.

The second is the safety standard or minimum quality standard. Here the purpose of the standard is slightly different. It is concerned with addressing questions such as the following. Is the product safe? Does it reach a minimum quality threshold? One object of this standard is simply to protect the consumer. More generally, the object of this type of standard is to overcome certain sorts of market failure that arise if the consumer cannot asses the quality of what (s)he is buying. These standards tend to be most effective if they are formal rather than informal.

The third is the standard to reduce variety. A familiar example of this is the clothing size. The objective of producing clothes in a limited number of standard sizes is to achieve economies of scale and economies of stockholding. Of course, there is also an element of compatibility standard here: the size 12 foot cannot fit into a size 8 shoe. These standard sizes can be formal or informal.

The fourth is the standard for measurement and description. These are, in a sense, a higher level of standard, because they don't so much refer to a standard for a particular product but rather a standard for the units we use to measure and describe the products we buy and sell. Obvious examples include standards of length, volume and weight. These standards tend to work best if they are formal, though some informal measurement standards also exist.

A STANDARDS 'RACE'

In this section we use the simple analytical approach presented above to describe the character of the 'standards race', in which competition between two or more competing product designs produces a 'dominant design' or 'de facto' standard. We then describe some of the strategies that firms use to increase their chances of winning the standards race.

An Analysis of the Standards Race[10]

We can use the framework above to illustrate the standards race at work. In the race, there are two products – both sold at the same price. One product we shall call the 'established standard' and the other we shall call the 'better replacement'. The first is of inferior quality, but it is an established product with a large existing network of users. The second is of superior quality, but it is a new product and has no network of users at the start of the race.

As illustrated in Figure 7.1 above, consumer choice between these two products will depend on the shape of consumer indifference curves between quality and network effects. In Figure 7.1 we drew two polar types of consumer indifference curve. At one end is the 'techie' who cares mainly about technological sophistication (or quality) and does not care much about network effects. At the other end is the 'networker' whose preferences are the other way round. In this illustration of the standards race, we assume that there is a continuum of customer types, ranging from the 'techie' to the 'networker'.

Let us imagine that we rank these customer types in order, according to the slope of their indifference curves, from 'techie' to 'networker', and let us identify the median customer. It turns out in races of this sort that the median customer is decisive to the outcome. If the median customer in a particular period prefers the established standard to the better replacement, then more than 50 per cent of customers in that period will choose the established standard. This means that the network of customers using the established standard will grow by more than the network of customers using the better replacement, and as a result the competitive position of the established standard will strengthen. Conversely, if the median customer in a particular period prefers the better replacement to the established standard, then more than 50 per cent of customers in that period will choose the better replacement. This means that the network of customers using the better replacement will grow by more than the network of customers using the established standard, and as a result the competitive position of the better replacement will strengthen. Finally, if the median customer is indifferent between the two products, then they each share 50 per cent of sales in that period, and the competitive balance between them is unchanged.

Figures 7.2, 7.3 and 7.4 below illustrate three outcomes to the standards race under some simplifying assumptions.[11] The first (Figure 7.2) is the 'neck and neck' race where the established standard (A) and the better replacement (B) continue to split the market equally between them, and neither wins the race. This outcome is theoretically possible but pretty unlikely. The second (Figure 7.3) is the case where the established standard (A) forges ahead because the replacement (B), better though it may be, is not good enough to make up for its smaller network of users. The third (Figure 7.4) is the case where the better replacement (B) *is* good enough to make up for its smaller network of users, and it catches up and eventually overtakes the established standard (A).

Each of these figures is in four parts. In the upper half of Figures 7.2, 7.3 and 7.4, we illustrate the choice facing the customer in period 1 (left) and period 6 (right), and we show the indifference curve of the median customer. In the lower half of each figure, we show the market share split in each period and the trend in the total network of users for products A and B. We see in Figure 7.2 that the market share split is always 50:50 and hence the total network sizes grow in parallel. In Figure 7.3, the market share for the established standard starts off above 50 per cent, and grows, and as a result the total network using the established standard forges ahead. In Figure 7.4, the market share for the better replacement is always above 50 per cent, and grows, so that the total network using the better replacement catches up and eventually overtakes the network for the established standard.

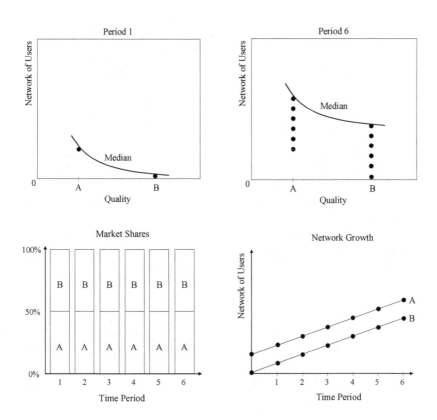

Figure 7.2   A 'neck and neck' standards race

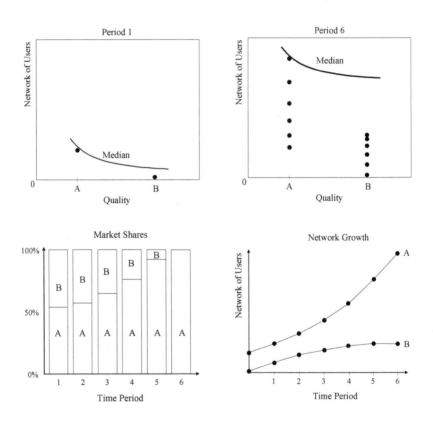

Figure 7.3   The established standard forges ahead

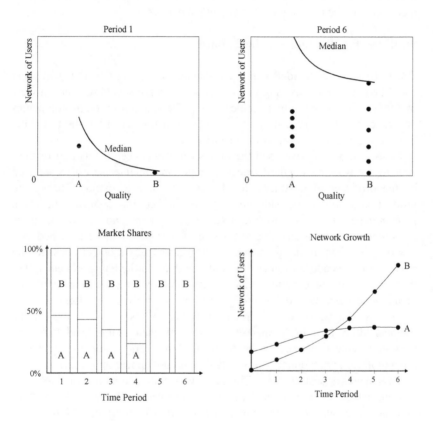

Figure 7.4   The better replacement catches up

This model of a standards race is very simple, but it captures the essential character of the race. Those who get ahead at the beginning of the race may stay ahead for the rest of the race. Small changes to strategy that may give one product a greater share of the market at the start of the race may be decisive in ensuring that product wins the race. This is a simple example of a more general phenomenon known as *path dependence*.

Strategies for Winning Standards Races

The last section concluded that small changes in strategy at a critical stage could be decisive in ensuring that a product wins the standards race. In view of that, it is important to understand some of the strategies used by companies in their efforts to win standards races.[12] In no particular order, the following is a list of some of the most common strategies.

Many companies have used the product preannouncement as part of their strategy for winning standards races. Product preannouncement is often used by the producer of a system that is relatively late to market. Before the product is ready to be launched on the market, the company makes a preannouncement (often to the trade press, or perhaps at an industry conference or exhibition): this states that their system will be launched in a few months' time. In the present context, the company's object is usually an attempt to persuade customers to wait for their forthcoming system rather than buy from the existing range of available systems. If successful, the preannouncement will delay the growth of the network of users of rival systems, so that when the company is ready to bring its product to market, the established products in the market do not enjoy such a large head start. The preannouncement has been open to abuse. First, some companies gained a reputation for preannouncing products that did not appear until much later, or perhaps never appeared at all. The latter was quite common in the software market, where it became known as 'vapour-ware'. In the light of this, preannouncements from some companies could lack credibility. Moreover, it was alleged (as part of an anti-trust case against IBM) that the preannouncement could be used in an anti-competitive fashion.[13]

A second strategy for winning standards races has been to recognise that 'the best is the enemy of the good'. To stand a chance of winning a standards race, the company cannot risk being too late to market. This may mean that it is better to bring the product to market as soon as it is 'good' rather than to wait until it is at its 'best'. Indeed, some have suggested that so intense is the pressure to bring software products to market promptly that companies may decide to market their products even if these software products may still contain 'bugs'.

A third, and related strategy, again common in the software market, is to sign up beta testers for a new software package. This is a way of releasing an early version of the software to selected high-value customers, of encouraging them to invest some time in learning to use the software. Such beta testers may help the software producer to identify faults or areas for improvement and that is of course very important. But in the context of a standards race, the use of beta testers may help to build up a network of users in advance of general release.

Several other strategies can be important in this context. One is the explicit sponsorship of some 'blue chip' customers, to ensure that these important companies become part of the network using a new product. Another is the strategy of building up indirect network effects by licensing third party producers to make a variety of add-on products that can be used in conjunction with your product. A further related strategy is to develop gateways from other products to your own, so that you reduce the user's costs of switching from a rival system to your own system. Finally, the VHS versus Betamax case study (see below) illustrates the importance of open standards as a way to build up a network of users of your technology.

STANDARDS RACES IN VIDEO RECORDING

There have been two quite distinct standards races in the history of video recording technology. The first is the famous market battle between VHS and Betamax in the mid-1970s to become the de facto standard for video cassette recorders. The second concerns the recent battle (2007) between Blu-ray and HD-DVD to become the de facto standard for high definition DVD.

The standards race between VHS and Betamax is one of the best known standards races of all. Betamax was the video cassette recorder system designed by Sony, a strong market leader in consumer electronics and renowned for its high quality. VHS was the system produced by the JVC company, a smaller company and less well known in consumer electronics. The two systems were incompatible with each other, using different shapes of cassette and a different electronic configuration. The two companies competed head to head for dominance of the VCR market in the mid-1970s. (A third system by Philips was also available, but was never one of the front runners.) Many observers expected that the Sony system would emerge as the de facto standard, but to the surprise of many, it did not.

One popular account of this standards race is that the original, basic system (VHS) was not overtaken by the later, superior technology (Betamax) because the former was quicker to build an extensive library of compatible software (films, recorded television programmes, etc). This is a nice story but is inaccurate in three respects: (a) Betamax was in fact the first system of the two to reach the market; (b) although Betamax was superior in some aspects of picture quality, VHS was superior in terms of the length of programmes that could be recorded; (c) the difference between the network effects (recorded software) for the two systems was not as great as this story suggests.

The historical record suggests instead that the key to the success of JVC and its VHS system lay in the fact that the VHS was a relatively open standard. JVC, being a small company, and knowing that they could not dominate this market on their own, were quick to license their technology to other consumer electronics manufacturers who could sell their own versions of the VHS video recorder. Sony, by contrast, made their Betamax closed standard. They did not license any other companies to produce versions of Betamax. It was this difference in standards strategy, and the relative sales effort behind the two standards that led the VHS system to forge ahead. Grindley (1995) discusses this case in more detail and gives other examples of the importance of adopting a relatively open standards strategy in standards races.

The VHS versus Betamax standards race was very expensive for the loser (Sony) who have always tried to keep a reputation of supporting their customers – even those who own old systems. This meant that Sony were much slower to concede that the race was lost than another company, more focused on profitability, might have been. The race was also expensive, though in a more modest way, for those consumers who initially bought the Betamax system, then found it was not to become the standard, and then bought a VHS system in addition. David (1985) has described these as the 'angry orphans' – a very vivid expression that needs no explanation.

Recognising that such market battles between standards can be costly for sellers and customers alike, many commentators on the emerging market for high definition DVD urged the competitors in that market to avoid another costly battle. But history would repeat itself!

There were two rival and incompatible systems competing for dominance in the high definition DVD market. One was the HD-DVD, backed by NEC, Toshiba and Microsoft, and the other was Blu-ray, from Sony and others. Technologically speaking, both have their advantages. The HD-DVD system was backwards compatible, so it is possible for users of that system to continue to play their existing collection of (conventional) DVDs. On the other hand, Blu-ray discs have greater capacity, which makes it possible to produce higher quality films and additional features.

Many commentators hoped that the two camps would recognise that a 'format war' is undesirable for all, and that they might declare a truce, thus avoiding a repeat of the VHS-Betamax battle. But it proved too difficult to broker such a compromise, and it became clear that the race could only be decided by market competition between the two systems.

HD-DVD had enjoyed a slightly earlier start in the market, slightly lower manufacturing costs and enjoys backwards compatibility. But, on the other hand, Blu-ray had more backers amongst the film studios and more companies on board to manufacture the system. By February 2008, when retail giant Wal-Mart announced that it would only stock Sony's Blu-ray format, and when Warner Bros and other DVD producers announced that they would only use the Blu-ray format, it was clear that, this time, Sony had won the standards battle.[14]

PUBLIC POLICY AND STANDARDS

The last section was about standards races and some of the strategies that firms may use in their efforts to win such races. It is obviously important to the firm in a standards race that it should win and not one of its rivals. But from a public policy point of view, it is arguably better that no proprietary product should 'win' a standards race, but rather that a public and *open* standard should emerge that all companies can use and produce.

Standards play an important role in technology policy for the following reasons. First, standards can help to reduce transaction standards and hence develop specialisation and the division of labour which, as we shall see below, plays an important role in increasing innovation and productivity. Second, genuinely open standards can enhance competition in open markets, whereas closed standards can act as a barrier to entry. Third, international standards support international trade and specialisation: we shall see a striking example of this in Chapter 14. Fourth, respected standards can help to reinforce trust between trading partners, which is another important issue in reducing transaction costs. Fifth, open standards play an important role in innovation, partly because of the reasons listed above, but also for other reasons: they make it easier for small scale entrants to sell innovative add-on products; they can help to diffuse best practice across firms; and they make it easier for producers and users to exploit network effects. A full discussion of these issues lies beyond the scope of this book, but for a discussion of the role of standards in supporting innovation, see Temple et al. (2005).

NOTES

[1] Some of the literature speaks of network *externalities* rather than network effects. Here we shall speak only of network effects, because the issue of whether such effects should or

should not be thought of as externalities is of secondary importance for the purposes of this book.

[2]   Casual observation suggests that there are also some negative network effects here: if a network is too open, it is no longer useful for communicating with a select group.

[3]   Indeed, I might prefer it if not too many others drive the same car as me, for there is no distinction in driving a commonplace brand.

[4]   See Laffont et al. (1998) for example.

[5]   Klemperer (1987).

[6]   The number of combinations of two callers is given by $N(N - 1)/2$. If each of these conversations is of equal value and of value to caller and receiver alike, then the total value is proportional to this number of combinations. For large N, this is roughly proportional to N^2.

[7]   To be precise, in a network of size N, the number of groups of size 3 and above is given by $2^N - N - 1$. For large N, this is roughly equal to 2^N.

[8]   Birke and Swann (2005) find that university students coordinate their choice of mobile phone operator to economise on call costs.

[9]   In the literature, the use of the term 'de jure' is incorrect because strictly speaking that implies that the standards have force in law. In general, we should use the word 'regulations' rather than 'standards' if these have force in law. By contrast, many standards are 'voluntary' in the sense that companies do not have to observe them by law, but choose to observe them because that makes good commercial sense.

[10] Katz and Shapiro (1985) and Farrell and Saloner (1985) provide rigorous analyses of the standards race.

[11] These assumptions include the following. We assume that total sales in each period are the same, and every customer buys one product or the other. We assume that there is no change in the slope of the median customer's indifference curve. We assume that there are no random factors influencing consumer choice in any period. We assume that customers have an unsophisticated way of forming expectations: they assume that difference in the sizes of the two networks in future will be the same absolute difference as it is when they purchase. And we assume, in effect, that the value of increased network size is proportional to network size, so there are no diminishing returns.

[12] Further details are provided by Grindley (1995) and Gabel (1991).

[13] This issue is discussed by Fisher et al. (1983).

[14] Guardian (2008).

8. Intellectual property

The inventions resulting from creative activity are often considered a form of property. As ideas, these are usually intangible and as such represent a rather special sort of property. It is customary to call this *intellectual property* (IP). This chapter looks at the special features of intellectual property.

We start with a discussion of why it if often necessary for companies to protect their intellectual property and examine the various methods used to protect IP. These can be divided into two categories: institutional methods and strategic methods. Institutional methods include patents, copyright, registered designs and trademarks. Strategic methods include commercial secrecy, confidentiality agreements, and the strategic use of complexity and lead-time to protect IP.

In the last two sections of this chapter we shall look at a fundamental dilemma in the protection of intellectual property. While it may be necessary to provide IP rights to provide enough incentive for creative activity, IP protection does in some ways run counter to the general economic principle that it is efficient to allow users to acquire IP at marginal cost. Since much intellectual property can be written down in a word-processed file on a computer, the marginal cost of copying is effectively zero. We conclude with a brief discussion of the open source movement, which strikes a balance between the two sides of this dilemma by maintaining well-defined IP rights but also allowing free copying and modification of the original sources.

Many texts on the economics of innovation devote a good deal more space to this topic than I do. The student who wishes to read further about intellectual property is referred to the supplementary reading on this topic in the Appendix.

WHAT IS INTELLECTUAL PROPERTY?

The term *intellectual property* (IP) refers to the output of a creative process such as an invention. The term entails three things. First, it is intellectual – that is, it is the product of the mind or the intellect. Second, as such it is often a very intangible asset. It may be represented in a particular form (e.g. a book is written on paper) but the value of this intellectual property is not limited by the particular physical form that it takes. Third, IP is treated in Law as property – just as *any other* property. This last observation opens up an interesting difference between the perspectives of the two disciplines, Law and Economics, on IP. In Law, IP is property like any other property. In Economics, however, we generally see intellectual property occupying an intermediate position between a free resource and tangible property. In Economics, IP is treated as property so long as it is useful to treat it in that way, but after some period of time, it may be treated more like a commonly owned good – for reasons that will be explored in more detail below.

WHY DOES IP NEED PROTECTION?

To a lawyer, the question in the title of this section would seem a very strange one! After all, IP is property and IP, like other forms of property, needs protection from theft or abuse. My family's car is our property and needs a lock to protect it from theft. A similar argument applies to IP.[1] Indeed, the lawyer might argue that IP needs particular forms of protection because theft is so easy. It is easy (but illegal) to make copies of a CD, or to copy musical tracks downloaded from the Internet. It is easy to photocopy a confidential document which describes a new invention or formula. In short, the marginal cost of copying someone else's IP is often very low indeed – especially so when the intellectual property can be *codified* (written down on paper or as a computer file).[2]

To the economist, however, the question posed in the title of this section is not such a strange one. To the economist, the reason for protecting IP is not so much because IP is property in some absolute sense. The reason has more to do with creating an economic environment in which creative firms have sufficient incentives to create their inventions. We said in Chapter 2 that invention was the result of a difficult (and often expensive) process of research and creativity. Most firms will only spend on this process if they are confident that they will be able to achieve some economic return on their investment. That requires two things. First, that there is a reasonable chance the invention will in due course be of commercial value. Second, that the inventor gets some kind of head start in using the invention for economic advantage. If at the end of this difficult and expensive process the inventor find that his/her invention can be copied easily by rivals, then that will usually undermine the economic return. And if the economic return is removed, then the incentive will disappear. For the economist, therefore, the rationale for protecting IP is instrumental: by protecting IP, we assure the creative firms that they will retain the economic value of their inventions and hence encourage them to invest their effort in creativity.

The fact that it is sometimes so easy and cheap to copy IP has several implications. First, some companies try to find clever ways to physically prevent any copying. For example, some software companies have introduced small 'patches' into their software so that it cannot be copied by any conventional means. Some music companies have done the same sort of thing with their CDs to prevent them from being copied on (say) a personal computer. And some consultancy firms place minute magnets in or on the pages of their consultancy reports so that they cannot be photocopied on a conventional photocopier, which becomes disoriented by the magnetic field. But in general it is impossible to come up with an entirely reliable method of physically preventing copying. That is why companies resort to legal and strategic measures to limit copying, as described in the next few sections.

Second, for the economist, the observation that it is cheap to copy IP poses an additional question: if it is so costless to copy intellectual property, are there then circumstances in which it would be economically efficient to allow such copying? The answer is that there are such circumstances. For example, the fair use arrangements for photocopying in libraries allow me to photocopy a *single chapter* of a book held without any direct payment to the author or his/her publisher. But fair use does not allow me to photocopy the entire book; if I want that, I have to *buy* a copy. This is a complex question to which we return at the end of this chapter.

METHODS OF IP PROTECTION: AN OVERVIEW

There are two broad approaches to protecting intellectual property. One is to obtain some official mark of recognition that the invention is the property of the inventor and may not be used for commercial advantage by others. The other is to keep the invention strictly secret – at least until the inventor is in a position to turn the invention into an innovation and maybe longer, in the case of a process invention.

The main formal methods of intellectual property protection are patents, registered designs, trademarks and copyright. There are important differences between these (discussed in the next section) but their economic effects are similar – at least to a first approximation. All four of them make the invention public but ensure (through the force of law) that no one else may use or copy the invention. If a rival does violate a patent, trademark, registered design or copyright, then the inventor may sue for civil damages. The effectiveness of these formal methods varies from country to country, however. They offer pretty strong protection in countries such as the USA and the UK but weaker protection in some other countries.

The most common informal methods of intellectual property protection are ensuring product/process complexity, wise use of lead-time, confidentiality agreements and other strategies for secrecy. Again, there are important differences between these (see below) but the basic idea is to keep the invention secret for as long as possible, or ensure that products and processes are so complex that even if some fragment of information were to be leaked, it would be of no value to a rival without all the other product/process details.

In practice, we find that the strategic methods are probably more important than the formal or institutional methods. Table 8.1 provides some evidence on this.

Table 8.1   Methods of protecting IP

Share of companies saying that method is very important for protecting IP

Lead-time advantage	30%
Confidentiality agreements	28%
Secrecy	28%
Complexity of design	24%
Trademarks	19%
Copyright	18%
Patents	13%
Registration of design	13%

Source: Author's calculations based on data in Community Innovation Survey (CIS3).

It is interesting to see that the four formal methods come in the last four places in the table, while the four strategic methods come at the top of the table. Why is this? Some companies have suggested (in interviews) that the patent system is expensive and time consuming to use and at the end of the process, the protection it offers is inadequate. From their point of view, it is better to use a strategic approach to protecting IP.

There are some variations by industry of course, with chemical and pharmaceuticals standing out as two industries where the patent is exceptionally important. There are also some important differences by size of firm, as shown in Table 8.2.

Table 8.2   Variations by company size in methods of protecting IP

Share of companies saying that method is very important for protecting IP	Employees		
	1-250	251-2500	2501+
Lead-time advantage	29%	58%	66%
Confidentiality agreements	27%	58%	78%
Secrecy	27%	57%	65%
Complexity of design	23%	49%	47%
Trademarks	18%	43%	54%
Copyright	17%	35%	57%
Patents	12%	35%	47%
Registration of design	12%	31%	28%

Source: Author's calculations based on data in Community Innovation Survey (CIS3).

Patent lawyers often argue that patents are much more important in large companies, and the table supports that assertion. But all types of IP protection (formal and strategic) are more heavily used by large companies than by the smallest companies. Whatever size of company we look at, the most important methods of IP protection remain the strategic ones (notably, confidentiality agreements, lead-time advantages and secrecy).

INSTITUTIONAL METHODS OF IP PROTECTION[3]

Patents

The patent gives a firm a monopoly right to commercial use of a particular invention (embodied in a product or process) for a given period. In the USA, this period is twenty years from the date of filing the patent. There are three key elements to the patent:

1) The patent grants a set of exclusive (or monopoly) rights to the inventor.
2) The patent holder has a temporary right to prevent others from making commercial use of the invention. The patent applies for a fixed term and after that, the patent-holder no longer has these monopoly rights.
3) In return, the patent holder must set out in the patent all the details of the invention, and this patent will be made available for public inspection.

In the absence of a patent, some inventions might be copied comparatively freely by many firms other than the originator, and the inventor would not recoup enough to cover the costs of his/her invention. In such a setting the incentive to invent would start to decline, and that would be a bad thing for the long-term prospects of the economy. The aim of the patent is to sustain the incentive to innovate. From an economist's point of view, this is more important than the sheer protection of intellectual property. When the originator has recouped his/her costs and a reasonable profit has been made, then, from the economist's perspective, the patent has served its purpose and could (or even *should*) lapse.[4]

Copyright

Copyright limits 'the right to copy' a piece of intellectual property. It applies to a wide range of creative works: poems, plays, novels, textbooks, journal articles, films, musical compositions and recordings, photographs, software, radio and television broadcasts, and so on. Copyright law covers *only* the particular form in which ideas have been expressed – not the underlying ideas. So, for example, the copyright of a Mickey Mouse cartoon prohibits anyone (except Disney) from making or distributing copies of the cartoon or from creating derivative works which mimic this particular mouse. But it does not prohibit people from drawing different cartoons about 'anthropomorphic mice' – so long as they are sufficiently different!

Registered Designs

A registered design refers to the appearance of the whole or a part of a product. The registered design gives the holder a monopoly right to use such a design in such a context. The registered design may refer to particular product features, in particular: lines, contours, colours, shape, texture, materials. The registered design cannot be concerned only with how a product works, or be concerned with designs for components of products that would not be visible in normal use. As with a patent, the design must be new and have an individual character. As with other methods of IP protection, registration is for a fixed period (up to 25 years). In law, a registered design is treated as property like any other business commodity. Specifically, it can be bought, sold, or licensed.

Trademarks

A trademark is a distinctive sign, mark or logo which distinguishes the goods and services of one company from those of another. The trademark may consist of a variety of different things: names, words, logos, symbols and pictures. The purpose of the trademark is to help the customer to quickly recognise that a particular product or service emanates from a particular company. To be an acceptable form of trademark for registration, the mark must be distinctive for the goods or services to which it is applied and must not be similar to any earlier marks for the same or similar goods or services. This means that if I register a particular trademark for the sale of computer software, there must be no other company using that same trademark for selling software. However, there may be another company using the same trademark for selling potatoes.[5]   The trademark must not deceive the customer by seeming to imply that the product is made by a company other than the one that actually made it. That is called 'passing off' and is a violation of the trademark holder's IP rights. Once again, the trademark is treated in law as a type of property, which can be bought, sold and licensed.

Case For and Against

All the mechanisms can help to provide creative companies with the assurance that they will be able to capture the economic value of their creative investments. These mechanisms all grant some rights to the creator of intellectual property in return for which the creator must put this intellectual property into the public domain. This latter condition is important, because it means that others may learn from the IP, which is economically efficient, but may not use it to compete against the creator of the IP. In this way, these mechanisms can overcome the free-riding problem and ensure that the creator achieves an adequate return from his/her creation. This may be essential to ensure that there are sufficient economic incentives for creative activity.

The main argument against these mechanisms is that they are all forms of monopoly and, on the whole, economics is strongly against monopoly. The justification of the temporary monopoly is that without it there are insufficient incentives for creative activity. But there seem to be many cases where those who own IP rights and have a monopoly in the use of an invention may make extraordinarily large windfall gains – far larger, surely, than were required to incentivise them to create the IP in the first case. In short, it seems that sometimes these mechanisms provide excessive protection and that is undesirable. A particularly undesirable use of the patent, for example, is for what is called 'pre-emptive patenting'. This is when those who hold a patent may not actually use the invention themselves, but may ensure that nobody else uses it either. This 'pre-emptive patenting' is highly undesirable because it means that the economy at large gains no benefit from the invention. (It may also be illegal in some countries.) Pre-emptive patenting in pharmaceuticals can be especially undesirable: it means that a potentially important invention, which could be used to create a treatment for an illness, cannot be used if the patent owner chooses not to use or to license it.[6]

STRATEGIC METHODS OF IP PROTECTION

Confidentiality Agreement

A confidentiality agreement or non-disclosure agreement is a legal contract between at least two parties in which they agree not to share some confidential materials with any *third party*.[7] The confidentiality agreement will list the confidential materials which the two parties can share with each other, but which they agree not to share with any others. Confidentiality agreements can (and indeed often *are* – see Tables 8.1 and 8.2) used to protect intellectual property. Whereas patents, copyrights, registered designs and trademarks *secure a monopoly right* to use the intellectual property in return for laying the IP open to public inspection, the confidentiality agreement keeps the IP secret. The confidentiality agreement aims to prevent others from using the intellectual property by keeping it secret between the parties to the confidentiality agreement.

The confidentiality agreement can protect IP that could in principle be protected by other means (e.g. patents) but can also protect IP that is hard or impossible to protect by any of the institutional methods.

Secrecy

Sometimes companies protect their intellectual property by keeping it secret from all outsiders and almost all insiders. In contrast to the confidentiality agreement which shares the IP with a limited number of others but binds them to confidentiality, the secrecy strategy is not to share IP at all. One of the most famous and closely-guarded examples of a trade secret is the Coca-Cola formula. The company says that only a few employees know or have access to the formula. While most of the constituent parts can be identified by food scientists, there is allegedly a 'secret ingredient' which has been a secret since the invention of Coca-Cola in 1886. It is said that the formula for this 'secret ingredient' is kept in a bank vault. Despite many attempts to 'reverse engineer' the 'secret ingredient', and a large literature on the secret formula, the Coca-Cola Company maintains that all published recipes are incorrect.

Lead-Time

A rather different approach to protecting the economic value of IP is to use it quickly before anyone else can. This is called the lead-time strategy. The first company to make commercial use of an invention can often earn a lead-time advantage. This means that so long as that company is the only player in the market, then it enjoys a temporary monopoly advantage – even if its intellectual property rapidly leaks out after it first enters the market. The first-mover advantages of lead-time are only temporary. The more successful the first entrant, the more tempting the invitation to competitors. The length of lead-time enjoyed by an innovator before competitors enter the market depends on several factors. It will depend on the nature of the innovation, the complexity of the production process and any other barriers to entry. For example, a competitor may take only a few days to copy some online business models, while it may take much longer for a competitor to duplicate complex manufacturing processes.

It is often argued – in the strategy literature, for example[8] – that even if first movers have a modest lead-time advantage over competitors, the fast seconds may be well placed to learn from the mistakes of the first mover, and may ultimately be the most successful players in the market. The following are examples of products where the 'fast second' did better than the first mover:

- Lotus 1-2-3 was more successful Visicalc (spreadsheet software)
- JVC's VHS system was more successful than Philips (video recorders)
- Internet Explorer eventually drove out Netscape (Internet browsers).

Complexity

Following on from the discussion of lead-time, it is generally recognised that the lead-time earned by the first entrant into a new market may be greater if the first entrant uses a new and highly complex process, which is hard to copy. This means that companies may be better able to protect their IP if it is only of value within a complex process that rivals do not currently use. This systematic cultivation of complexity to secure the economic value from IP will work even better when some of the other key knowledge required to exploit the IP is *tacit* (held in the minds of employees) and not *codified* (written down on paper).

Case For and Against

The cases for and against these strategic methods of IP protection are essentially the mirror image of the cases for and against formal methods of IP protection. As we said before, in general economics thinks that monopoly is a bad thing, and we tend to think that private monopolies protected by public policy are doubly bad. The advantage of these strategic methods is that they do not create any publicly protected monopolies and hence the problems discussed above when commenting on the case against formal methods do not apply.

However, the reliance on confidentiality and secrecy means that we miss one of the greatest benefits of the patent system (and other forms of institutional IP protection). IP remains secret and is not available for public inspection. This means that an important channel by which firms learn from each other is closed.

A FUNDAMENTAL DILEMMA IN IP PROTECTION

We have noted above that economics recognises a fundamental dilemma about the protection of intellectual property. Economics recognises that we need to maintain incentives if we are to encourage companies to go to the expense and risk of creating IP. If my investment creates IP that can be immediately used by all my rivals (who have made no contribution to the cost) and gives me no advantage over them, then I may wonder if the investment is worthwhile. On the other hand, it is economically efficient to diffuse use of IP at marginal cost – and with IP that is often close to zero, for the reasons discussed above. Our best attempt at resolving this dilemma is to limit the protection offered by the formal methods. These limits mean that formal methods of protection restrict others from using the IP during a particular time period and in a particular sub-set of economic activities. In the context of patents, we restrict the scope (range of applications) and length (time period) of the patent and similar restrictions apply to some other mechanisms.

We said above that from the point of view of Law, IP is property – just like any other property. From the point of view of Economics, the perspective is a bit different. To provide inventors with the incentive to create new inventions, we allow them to treat these as their own property for a while. But if the cost of copying this intellectual property to others is very low, then it is economically inefficient to allow this intellectual property to be monopolised. This presents the economist with a subtle trade-off: we need to maintain incentives but we don't want to hold up the diffusion of new inventions. To achieve the right balance, patent agencies (and other agencies that protect IP) adjust the length of the monopoly and its scope. When the patent expires then others may copy the IP at marginal cost.

OPEN SOURCE

We conclude this chapter with some brief observations about a recent trend in IP protection (or non-protection) which, some would say, is an obvious consequence of the dilemma described in the last section.

In software, an *open source* community has arisen. In particular we shall focus on what have become known as Free (*Libre*) Open Source Software (FLOSS) projects.[9] In this movement, companies share some of the intellectual property with others even if that means that they lose some of the potential economic value of their IP by sharing in this way. Open source (as defined in the FLOSS projects) means three things:

1) Intellectual property rights are not abandoned. The developers of intellectual property retain rights over their property but allow others to use it too. To repeat what we said in Chapter 2, it is a little bit like the country landowner who allows people to walk on his/her land so long as they do no damage. This action does not make the land 'common land': the landowner is still definitely the owner. But (s)he does not assert a monopoly over the use of the land for recreation.

2) The original source code of the software is made open to other software developers. This is in contrast to the norm with much PC industry software in which only the companies developing a particular operating system or applications package have access to the source code.

3) Software can be freely redistributed. This means that users of the software are licensed to give away the software (as part of a collection of software from different sources) and no royalty or other fee needs to be paid to do this.

What is the point of this? We can answer this question at two levels. Why does open source create better software? Why would companies want to do this?

First, why does open source create better software? The Open Source website[10] puts the case very well:

> The basic idea behind open source is very simple: When programmers can read, redistribute, and modify the source code for a piece of software, the software evolves ... We in the open source community have learned that this rapid evolutionary process produces better software than the traditional closed model, in

which only a very few programmers can see the source and everybody else must blindly use an opaque block of bits.

A useful article on Wikipedia[11] describes some of the wider economic benefits as follows:

> The idea of open source is then to eliminate the access costs of the consumer and the creator by reducing the restrictions of copyright. This will lead to creation of additional works, which build upon previous work and add to greater social benefit. Additionally, some proponents argue that open source also relieves society of the administration and enforcement costs of copyright.

But why would companies would want to give away some of the potential economic value of their IP in this way? One argument, due to Lerner and Tirole (2002) is that individual employees gift some of their IP (in the form of software) to a FLOSS project in order to gain professional recognition and thus to enhance their future career. In that respect, it is a little like academic publishing: we do it to raise our reputation and not because we get paid for it. But as Dalle and David (2007) have shown, the motivation for participation in open source projects is more diverse than this. Dalle and David group the developers into four types:

1) *Lerner-Tiroleans* – kudos-seeking programmers, attracted to work on projects for career reasons.
2) *Social Hackers* – who seek community recognition and esteem.
3) *Social Learners* – novices who seek community interaction for self-improvement.
4) *von Hippelites* – individualist 'user-innovators' who seek the opportunity to build some particular functional capability that they will find useful.

Dalle and David (2007) argue that successful operation of open source projects requires the participation of *all* these different sorts.

NOTES

[1] However, there is a difference between theft of tangible property and theft of intangible property like IP. If our car is stolen, my family and I no longer have a car to drive. If my intellectual property is stolen then I have not lost the property in its entirety. What I have lost is some (or all) of the economic value of that property. If someone makes illicit copies of one of my books, then I have not lost the book or the words, but I have lost the royalties I would have made from selling legitimate copies.

[2] Copying IP is much harder when the IP is tacit (contained in the mind of the inventor but not written down).

3   This section draws on material on the website of the UK Intellectual Property Office, http://www.ipo.gov.uk/

4   There is a huge (and not very edifying) literature on the optimum breadth and length of a patent.

5   So, for example, several companies use 'ABC' as a trademark, but in each case for selling different goods and services.

6   The more recent literature on patenting recognises various other abuses of the patent system. Increasingly, it is alleged, companies use patents not so much to protect valuable IP, but in order to create a bargaining stake for negotiations with other companies, and to act as a form of entry deterrent to small scale rivals. These issues are discussed by Hall (2005).

7   This legal language means that companies A and B agree between themselves that they will not share certain confidential information with companies C, D, E, etc.

8   See Geroski and Markides (2004).

9   http://www.opensource.org/

10   http://www.opensource.org/

11   http://en.wikipedia.org/wiki/Open_source

PART III

How firms achieve innovation

9. Invention and creativity

Creativity and invention are not the same as innovation but are essential inputs to the innovative process. In this chapter we review some of the key ideas in the theory of creativity and invention. Most of this literature originates from organisational psychology and individual psychology, and is rather different in character from the economics with which the student will be familiar. But we shall start with a brief review of what economics has to say about the processes that generate invention and creativity.

ECONOMICS AND INVENTION

The economics literature tends to talk of invention rather than creativity. That literature on invention lies at the confluence of two different streams of economic thought: the economics of networks and the division of labour.

Simon (1985) recognised that the process of learning from *diverse knowledge bases* is a highly important source of invention and innovation. To a first approximation, the larger the network of people from whom we can learn, the greater the prospects for invention. We have already discussed some of the key ideas about the economics of networks in Chapter 6. These principles seek to explain how the value of a network increases with size and composition and these arguments are relevant to the sort of learning process that Simon had in mind.

Chapter 14 will discuss in more detail the connection between the innovation and the division of labour. But the key insight was that by Adam Smith: he attributed much of the invention he observed to prior division of labour (Smith, 1776/1904a, p. 11):

> the invention of all those machines by which labour is so much facilitated and abridged seems to have been originally owing to the division of labour. Men are much more likely to discover easier and readier methods of attaining any object when the whole attention of their minds is directed towards that single object than when it is dissipated among a great variety of things ... A great part of the machines made use of in those manufactures in which labour is most subdivided, were originally the inventions of common workmen, who, being each of them employed in some very simple operation, naturally turned their thoughts towards finding out easier and readier methods of performing it.

From this perspective, invention and innovation are not activities that call for much networking. Specialised labour builds up enough experience through learning by doing from which to create inventions as a problem-solving exercise. The process of the division of labour creates an incentive for specialised labour to seek to modify their tools and invent new ones.

These two explanations of invention look very different indeed and may seem incompatible. But one of the main contributions of this chapter is to show that they are not incompatible. For there must be a limit to the extent of invention derived from the division of labour. Smith himself knew that the division of labour was not an unambiguously good thing for the economy and society. He described how those who spent all their lives performing a few operations would have very limited knowledge or intelligence, and their intellectual capital is bound to degrade. Most significantly, for our present purpose he also conceded that (Smith, 1776/1904b, p. 267; my emphasis):

> The man whose whole life is spent in performing a few simple operations, of which the effects are perhaps always the same, or very nearly the same, has no occasion to exert his understanding or to exercise his *invention*.

In short, while the division of labour may be a source of inventions and innovations, that source may dry up when the labour becomes too highly divided. What happens next? If divided labour does not have within itself the intelligence for invention, then that invention must be put together from different sources. Smith (1776/1904a, p. 12) recognised that some inventions were made by:

> those who are called philosophers or men of speculation, whose trade it is not to do anything, but to observe everything; and who, upon that account, are often capable of combining together the powers of the most distant and dissimilar objects.

This extraordinarily farsighted passage captures three essential characteristics of the professional 'man of speculation' (or inventor). First, his/her trade is 'not to do anything'. (S)he is a theorist. Second, (s)he observes everything. To do that (s)he must talk to many. Now, surely, invention and innovation become much more of a network activity. Third, (s)he is good at 'combining together' disparate and dissimilar knowledge. This leads us on to the second key perspective on the process of invention and innovation.

That is the essence of what the economics literature has to say about the two streams leading to invention and creativity. But we can learn a lot more by reading the creativity literature from psychology, and we turn to that now.

CREATIVITY: A PARADOX

As this is a book for economics students who are unlikely to want to follow up detailed references to the creativity literature, I have not given extensive references throughout this chapter, but have listed a few items of supplementary reading in the Appendix.

The creativity literature recognises a deep paradox about the creative process. One facet of this is that creativity requires both introversion and extroversion. The creative person very often has (and probably *must have*) personal characteristics correlated with introversion – notably, *autonomy*, a key concept in creativity. But the creative person cannot exist in a vacuum, and requires to connect with the rest of the world – both for ideas and for an audience. Another facet of this paradox is that creativity requires a delicate balance between obedience and disobedience. The creative person must question norms and disobey those norms that stifle his/her thought. At the same time, those who rebel face disapproval, criticism and isolation. There are 'rules about breaking the rules'.

The psychoanalyst, Otto Rank, recognised the fundamental tension here. Creative work stems from two desires that are in tension with each other: the desire for *individuation*, the ability to develop one's own autonomous self, and the desire for *identification* to share experiences and togetherness. Rank indeed recognised that the tension between these two is such that the transition towards individuation is painful. He recognised three types or stages of personality:

- A *conformist* or *adapted* type: these people have not developed their autonomy, and take their lead from the world around them. They passively obey norms and dare not move out of line.
- A *conflicted* or *neurotic* type: these people have moved some way towards developing their autonomy. They have broken free of some norms, but feel uneasy about this. This unease makes them unhappy and confused, and they spend much of their energy in a fight against external domination.
- A *creative* or *productive* type: these people have completed their passage through the two previous stages, and have emerged with a powerful autonomous voice. Instead of being engaged in a fight against domination, these people recognise and affirm themselves.

While Rank's categories apply to exceptional creativity – rather than everyday creativity – the creativity literature still seems to recognise such tensions for anyone involved in creativity, at whatever level.

BISOCIATION

The creativity literature sees combination and reorganisation as fundamental to the process of creative thought. People create new knowledge or ideas by combining and reorganising *existing* concepts or categories. This has been recognised for some time (if in an informal way) in the aphorisms of the great minds. Einstein, Feynman and others are all associated with the idea that creativity/invention, 'is seeing what everyone else has seen, and thinking what no one else has thought'. And indeed, the same idea is recognised in the ideas of Herbert Simon, noted at the start of this chapter.

Koestler (1964) coined the term *bisociation* to describe what happens in creative thinking. Koestler's aim was to, 'make a distinction between the routine skills of thinking on a single "plane" … and the creative act which … always operates on more than one plane.' Bisociation is about perceiving an idea or situation, 'in two self-consistent but habitually incompatible frames of reference.' (Koestler, 1964, p. 35)

Bisociation is a *combinatorial* activity – meaning that it involves the combination of existing ideas. It need not be a social activity, however, and we shall come back to that below. However, some have argued that the scope for bisociation is greatest when there can be creative interaction in heterogeneous groups. The economist is therefore tempted to explore, without more ado, whether Metcalfe's Law could be relevant to bisociation. And indeed, if networks enabled the frictionless bisociation of different frames of reference or knowledge bases, then Metcalfe's Law could indeed be useful. But the friction in this process should not be underestimated!

The literature on group creativity starts from the position that creative interaction in heterogeneous groups should help the creative process. Group interaction is important because it brings together individuals with different experience and backgrounds to exchange ideas. The more diverse the group, the greater the potential for creative bisociation because the group can in principle combine many different knowledge sets. This becomes increasingly important with the ongoing division of intellectual labour, and the attendant growing complexity of disciplines, because that makes it hard for any one individual to master more than one, or at best a very small number of disciplines.

However, the creativity literature doesn't stop there. It *recognises* the friction we described above. First, different disciplines generally lack a common language or common concepts, so that exchange between different members may be at a low level. Second, if the group progresses beyond the point of non-communication, it may well transpire that individuals from different backgrounds have quite different values, and hence disagreement or even conflict is quite possible.[1]

In theory such conflict can be productive, if managed properly.[2] But if not managed properly, it is liable to inhibit the creative process. In those exceptional cases where outstanding creativity has come from groups with very poor interpersonal relationships, it seems more likely that the creativity is achieved *despite* rather than *because of* the conflict.

For those groups that cannot cope with such difficult inter-personal relations, diplomatic silence may be in order. But then there is a risk that the group will fail to achieve the creative fusion that it might. It is recognised that even if careful principles are followed for group brainstorming, these groups may still generate fewer ideas than would emerge if the group members brainstorm in isolation. This may happen for a variety of reasons, but notably if the group dynamic and fear of ridicule makes members reluctant to toss half-formed ideas into the discussion.

Janis (1972, 1982) has coined the term, *Groupthink*, to describe the risks that a group may encounter in such circumstances. Groups that are unduly concerned to avoid conflict and achieve unanimity will often fail to explore all alternatives. Such groups may not air differences, may suppress dissent, may not seek expert advice and may tend to stereotype experts in an adverse way, may be too selective in the information they gather, and may have an illusion of invulnerability. As a result, groups suffering from *Groupthink* are liable to make poor decisions, which can sometimes be catastrophic.

In short, creative collaboration between diverse parties is liable to encounter friction, in one form or another. Sometimes the benefits of collaboration are enough to overcome the friction. Sometimes the friction is just too great.

As we said above, however, bisociation *need not be* an especially social activity. In principle, one scholar can achieve bisociation on his/her own. When one scholar masters two of the disciplines that one would combine in a group creativity exercise, then (s)he need not suffer the communication problems or the conflict that could arise in the group, though (s)he may suffer from a degree of cognitive dissonance.

A *hybrid* scholar is a researcher who transgresses the accepted boundaries of his/her home discipline and integrates concepts, theories, methods and results originating from other disciplines (Dogan and Pahre, 1990). Disciplines vary in the reaction to hybrid scholars: some are highly suspicious and often very hostile towards scholars who have travelled to other 'lands' and seek to return.[3]

The hybridisation of disciplines means that elements from overlapping or adjacent disciplines are recombined into new specialised fields. Dogan (1994) goes further to argue that much of the invention in each discipline depends largely on exchanges with other fields. From this point of view, the *marginal scholars* (located at the boundaries of their discipline and adjacent disciplines) play an especially important role in intellectual invention. Indeed Dogan and Pahre (1990) write of 'creative marginality', and go as far as to suggest that progress in academic disciplines is concentrated at the periphery, where there is cross-fertilisation with other disciplines. By contrast the core of the discipline becomes stagnant.

As a result, some have written of a 'paradox of density', which argues that in densely populated core fields, productivity per head is lower than in marginal fields. In core fields, a large part of the really innovative work is completed before the field becomes densely populated. Overcrowding simply encourages the spread of 'opaque jargons', intense debates about niceties, and safe but routine work carried out by the large community of intellectual *foot soldiers*.

AUTONOMY

So far we have spoken of one essential characteristic of creative work – bisociation, or the bringing together of different ideas, bodies of knowledge or frames of reference. Now we turn to another essential characteristic of creative work – the need for the creative person to establish his/her own intellectual and creative autonomy.

Research on creativity has identified several characteristics of personality that are regularly correlated with creativity. In summary, these include:

- Introversion
- Self-directedness and self-sufficiency
- Independence of mind and judgement
- Stubbornness and arrogance
- Courage in the face of criticism
- Intrinsic (rather than extrinsic) motivation
- Lack of concern about others' perceptions
- Little need for external approval
- No desire to conform for the sake of it
- Lack of interpersonal skills
- Asociability and even anti-sociability
- Liking for solitude.

We can perhaps summarise these in the term, *autonomy*. Creative people are either autonomous by nature (or have it forced on them), or have to create such autonomy. The creativity literature recognises that it is probably impossible to identify a single direction of causation here. Naturally autonomous people tend to be creative, but equally, creative work also tends to make people autonomous.

Sheldon (1999) argues that 'conformity and creativity don't mix'. Creativity of its very character involves breaking rules and disobeying norms. Sheldon argues that a large amount of research evidence shows that pressures to conform, broadly defined, have negative effects on creative effort. Sheldon makes a distinction between *informational* and *normative* social influences. *Informational* influence should be constructive if people use the information gained to sharpen their perception. In contrast, *normative* social influence can be destructive to the extent that it deters the creative person from his/her creative quest, back towards the conventional, flawed view.

Those with a well-developed sense of autonomy are better able to take all external influence as informational – whether the intention was informational or normative. Because autonomous people are not too concerned about winning the approval of those with whom they converse, they are better able to select what is helpful from the advice and criticism they receive and ignore the rest. Those who have not reached that state of autonomy, and are unduly concerned to win approval of their peers, will not find it easy to interpret all external influence in this way. Faced with normative influence that is negative about their present work, their anxiety level rises sharply, and they tend to hurry back to the 'straight and narrow' of conventional wisdom. Using Rank's three categories, the *conformist* type quickly alters his/her behaviour in the face of negative normative influence, while the productive type takes it in his/her stride, and just picks out what is useful. The *conflicted* type reacts unpredictably, even neurotically.

CONTRARIANISM AND ASYNCHRONY

Contrarianism is an extreme way in which the creative person may assert his/her autonomy. The contrarian automatically takes a position in opposition to current norms and conventions. The creativity literature is interested in the contrarian, but agrees that contrarianism cannot provide any guarantee of creativity. For it can be argued that the contrarian is arguably just as much the slave of convention as the adapted type. While the fully autonomous creative type does not allow normative influence to divert him/her from finding 'the truth', the opinions of adapted types and contrarians (though in perfect inversion to each other) are fully determined by current norms and conventions. The contrarian is bound to disregard norms even when they are right.

There is a rough linkage here to another contrary strategy, though not strictly contrarianism. Gardner and Wolf (1988) argue that periods of outstanding creativity tend to coincide with periods of 'asynchrony' or tension in the life and work of the creative person. Gardner and Wolf go on beyond this to argue that this creativity happens not *in spite of* the 'asynchrony' but *because of* it. In this vein, Hudson (1966) found that 'crisis seeking' is characteristic of creative thinkers and Barron (1963) found that scientists seeking originality were attracted to (rather than repelled by) disorder.

INTRINSIC AND EXTRINSIC MOTIVATION

Amabile (1996) has stressed what she considers a key distinction between *intrinsic* and *extrinsic* motivation for creative work. Researchers who are intrinsically motivated are creative because of their love or fascination for the subject and for what they create. Researchers who are extrinsically motivated are creative as a means to a rather different end – advancement, promotion, fame, peer recognition, and so on.

Amabile argues that extrinsic motivation is not as good for creativity as intrinsic motivation. To the extent that creativity involves breaking rules and rejecting norms and risking the wrath of the academy, the extrinsically motivated person may pull up shy when such creative work risks damaging his/her reputation, and will instead seek an accommodation with current norms. The intrinsically motivated researcher, by contrast, is much less worried about incurring the wrath of the academy. As a result (s)he is more task oriented, and will espouse unpopular and contrary views if (s)he believes they are right.

Subsequent writers have argued that extrinsic motivation is not necessarily opposed to creativity. At times, extrinsic motivation may supply a valuable complement to intrinsic motivation, and may even be stronger than intrinsic motivation. But there is general agreement that those who are predominantly influenced by extrinsic factors rather than intrinsic factors are unlikely to achieve exceptional creativity.

INCUBATION AND 'EPIPHANY'

One final area of the creativity literature is important in the context of this chapter. Several writers have shown that creativity usually does not take the form of sudden flashes of inspiration *out of the blue*. Rather, creativity is the culmination of long periods of sustained thought and effort.

This is not to deny that the final breakthrough may appear suddenly, but to emphasise that that breakthrough builds on long periods of painstaking thought. Everyone knows the story of Archimedes jumping from his bath shouting 'Eureka'; what we know less well is that this was a solution to a problem that had frustrated him for a long time. The literature uses the term *incubation* to describe how the creative person needs long periods of heavy concentration on a problem followed by periods detached from that problem, in order to allow the creative product or solution to emerge. Archimedes' shout of 'Eureka' followed a process of incubation.

Some of the creativity literature uses the word 'epiphany' to describe how several fruits of creative labour all 'ripen' at the same time. The creative person endures a long and sometimes painful process developing knowledge, emotions and goals, and at the end of this process the creative product emerges.

THE PARADOX RESOLVED

We started this chapter with a paradox. Creativity requires both introversion and extroversion and yet it is hard to find these in the same person. The phase model of creativity offers a partial resolution to the paradox. Yes, creativity requires sociable networking and introverted autonomy – *but not at the same time*.

The phase model recognises five stages to the process of creativity, and recognises that each stage calls for different activities. These are:

1) information
2) incubation
3) illumination
4) verification
5) communication.

The first phase (*information* gathering) is a relatively extroverted activity. At the very least, the creative thinker has to refer to what is known in the literature (information). By contrast, the second, third and fourth phases are relatively introverted. Starting from what is known already, the creative thinker starts a (possibly quite long) stage of *incubation*. With luck and a lot of hard work, that eventually leads to *illumination* – the 'Eureka' of Archimedes. But before that is communicated to any audience, the thinker must carry out some preliminary in-house tests to verify that the creative product works (*verification*). Only when these three relatively introverted phases are complete, would the creative thinker emerge and start the altogether more extroverted process of *communicating* creative ideas to an audience.

Some empirical studies of creativity have cast doubt on the empirical accuracy of this simple phase model. For one thing, in its simplest form, it is rather 'linear' or uni-directional in character. Nevertheless, it is a useful way to disentangle the various activities involved in creativity. It can perhaps be made a little more realistic by recognising that it contains various feedback loops. For example, adverse reaction from an audience is a form of information and may therefore feed back into the *information* set and start a fresh round creative thinking. Equally, adverse results from the thinker's own *verification* stage can lead to another period of *incubation*.

CONSISTENCY OF ECONOMICS AND CREATIVITY

Are these ideas from the creativity literature consistent with economists' ideas about invention? Yes they are, up to a point, but the ideas from creativity are rather more subtle. Let us look at the ideas above from the two key economic perspectives on invention and creativity.

Networks

Creativity is a combinatorial activity and the greater and more diverse the community who can be joined together in a network, the greater the theoretical benefits for creativity. However, as we have seen, the creativity literature recognises some of the frictions that emerge in group creativity, which economics tends to overlook. These frictions may well make it impossible to achieve these theoretical benefits.

We have also seen that exceptional creativity is correlated with certain personal characteristics that fit uneasily in the network, and will indeed be a source of such frictions. The creative person is autonomous, a norm doubter, is not unduly concerned about peer approval, shuns conformity for the sake of it, and is not unduly motivated by extrinsic factors. Such people do not feel an especial wish to form part of a network, and others may with justification feel unease or even irritation at having such people in their community.

We have also seen what can happen if the autonomous creative person feels under pressure to modify his/her behaviour to fit in with the group. Such normative influence can deaden his/her perception and although it may make the group work better at a superficial level it can also lead to a poorer group outcome, because important knowledge located within the group is not shared for diplomatic reasons. We encountered the phenomenon of *Groupthink* that is perhaps an extreme case of this, where group decisions can be dangerously flawed.

In short, the idea in the economics of networks (for example, 'Metcalfe's Law') that the value of a network increases with the number and diversity of participants, seems much too simplistic. It is a theoretical possibility but does not reckon with the frictions that can emerge.

Division of Labour

The fit between the creativity literature and the concept of division of labour is in part a close one and in part not so close. To understand this point it is helpful to make a distinction between containment and autonomy.

When there is division of labour, the activities of any person within that are *contained*. Tasks are well defined and specified in advance. Channels of communication are limited and predictable. Contrast this with the network firm or network trader whose tasks and channels of communication are *not* contained. Those within a division of labour are indeed, as Smith (1776/1904a, p. 11) put it, 'much more likely to discover easier and readier methods of attaining any object when the whole attention of their minds is directed towards that single object'.

However, we could not say that those within a division of labour enjoy *autonomy*. They are, in a manner of speaking, cogs in a greater machine. Their survival depends on their seamless integration into that machine. Autonomous behaviour towards the rest of the machine is not sustainable.

The discussion of personal characteristics correlated with creativity identified some that are consistent with a division of labour: introversion, lack of interpersonal skills, asociability and even anti-sociability, and perhaps a liking for solitude. These characteristics require containment but do not require autonomy. But some of these characteristics associated with creativity definitely require autonomy: self-directedness and self-sufficiency, independence of mind and judgement, stubbornness and arrogance, intrinsic (rather than extrinsic) motivation, lack of concern about others' perceptions, little need for external approval, and no desire to conform for the sake of it. This is not consistent with a division of labour.

Moving beyond personal characteristics, however, some other factors recognised in the creativity literature are also consistent with the division of labour. In particular, the idea of incubation recognises that the creative person needs long periods of careful attention to the same problem if (s)he is to produce a creative solution.

NOTES

1. An interesting example of this is the exchange between physicists and economists in the early days of the Santa Fe Institute (Waldrop, 1994). The physicists, for whom founding assumptions on fact is of paramount importance, were surprised at some of the unrealistic assumptions which the economic theorists were in the habit of making. The economic theorists, for whom theoretical and mathematical rigour was of paramount importance, were surprised that the physicists' approach to mathematics was comparatively casual.
2. Francis Crick, who shared the 1962 Nobel Prize for the discovery of DNA, is quoted as saying that politeness is the enemy of effective collaboration, which requires candour or even rudeness (Abra and Abra, 1999).
3. Scitovsky (1976), an outstanding and highly respected economist, tells of the uniform hostility that he met from all types of economist when he tried to develop a hybrid economic/psychological theory of consumption.

10. The entrepreneur and innovation

The concept of an entrepreneur is a very important one in economics, but in this brief chapter, we do not attempt to give an overview of all that is done by the entrepreneur. Rather, our aim is simply to explore what economic theory has to say about the contribution of the entrepreneur *to innovation.*

I have two reasons for keeping this chapter so brief. The first is that Earl and Wakeley (2005, Chapter 3) have already provided an excellent overview of the economic theory of the entrepreneur and I would refer the student wanting a broader overview to that chapter. The second is that while some authors suggest that innovation and entrepreneurship are inextricably linked, the reality is that much of what the entrepreneur does is not innovation and much of what the innovator does is not entrepreneurship. I focus here only on the intersection of these two categories.

In outline, Earl and Wakeley identify six different perspectives on what entrepreneurs do that distinguishes them from other economic actors.

- **Mainstream Neoclassical Economics:** Neoclassical economics does not have an especially compelling theory of the entrepreneur. The entrepreneur is just another factor of production (like labour). The contribution of the entrepreneur to the business is nothing unique. Many heterodox economists consider that this approach does not do justice to the entrepreneur.
- **Leibenstein:** From this perspective, the entrepreneur's role is to enhance the performance of inefficient firms, by reducing x-inefficiency.
- **Austrian Economics (Hayek, Kirzner):** From this perspective the entrepreneur is a force for equilibrium. The entrepreneur sees profit opportunities when a market is out of equilibrium. For example, where there is excess demand, the entrepreneur can make a profit by supplying that excess demand. In doing so, the entrepreneur brings the market back towards equilibrium.
- **Schumpeter:** For Schumpeter, the entrepreneur is an innovator. Indeed, any business activity that does not count as innovation would not count as entrepreneurship from Schumpeter's perspective.
- **Casson:** This perspective focuses on some of the distinctive things that

entrepreneurs do. In particular, Casson sees the entrepreneur as a specialist in coordination.

- **Shackle/Earl:** This is an unusual and very interesting perspective. Shackle's perspective on the entrepreneur (further developed by Earl, 2003) sees the entrepreneur as an experimenter who relishes in making unexplored connections.

For our present purposes, however, two are of particular importance to the study of innovation: Schumpeter's perspective and Shackle's perspective. We shall discuss these in turn.

SCHUMPETER: THE ENTREPRENEUR AS INNOVATOR

For Schumpeter the entrepreneur is more or less equivalent to the innovator. This stands in interesting contrast to the Austrian perspectives of Hayek and Kirzner (see Kirzner, 1979). Whereas the Austrian economists saw the entrepreneur as a force that brought about equilibrium, Schumpeter saw the entrepreneur having – in a sense – the precise opposite effect. The entrepreneur is a destroyer of equilibrium situations, and thinks up ways of putting scarce resources to new uses by introducing new goods or a new quality of goods, by introducing new ways of producing goods, by opening up new markets, by discovering new sources of supply of raw materials or partly manufactured goods, and by reorganising the structure of an industry (e.g. by creating a monopoly or breaking up a monopoly situation).

While Schumpeter's perspective is a powerful one, it has some slightly anomalous features. Recall our definition of innovation from Chapter 3. This was indeed Schumpeter's definition: innovation is the *first* commercial application of what up to that point has remained non-commercialised knowledge. From Schumpeter's perspective, as the entrepreneur is an innovator, then the entrepreneur must also be the first person to innovate. From that perspective, the second person into the market cannot be an entrepreneur: the second and subsequent entrants are simply imitators.

This seems to deny some important activity that would count as entrepreneurship from other perspectives. Moreover, for Schumpeter, the activity of running and managing the business after innovation is not entrepreneurship either: it is the more routine job of business administration. Again that seems to deny some important activity that would count as entrepreneurship from other perspectives. Finally, Schumpeter draws a clear distinction between entrepreneurs and capitalists: capitalists provide finance but entrepreneurs do not bear the financial risks associated with their innovations.

Most economists would agree, I think, that Schumpeter's description of the entrepreneur is too narrow. Nonetheless, for those interested in innovation, it is the ultimate *innovation-centred* description of the entrepreneur. For, in Schumpeter's view, if a business activity is not innovation, then it is not entrepreneurship either.

SHACKLE: THE ENTREPRENEUR AS A CONSTRUCTOR OF CONNECTIONS[1]

Earl (2003) calls this the 'Shackle' perspective on the entrepreneur, but in my view, Earl himself should take some of the credit for developing this view of the entrepreneur.

This perspective starts from the assumption that most new ideas are based upon a limited set of elements: these elements are combined in new ways to create new ideas. That assumption is similar to Koestler's (1964) concept of creative thinking as bisociation (see Chapter 9). In the Shackle/Earl theory, what eventually turns out to be a profit opportunity is initially conceived as a possibility in the mind of the entrepreneur. Profit opportunities are not things that lie around waiting to be found; the entrepreneur has to construct them actively. Entrepreneurs 'imagine what is deemed to be possible' (Shackle, here quoted from Earl, 2003) and this imagination involves the entrepreneur in recognising interesting connections between hitherto unconnected elements.

Shackle and Earl consider that entrepreneurs have certain characteristics and qualities which make them different from the rest of us. Such entrepreneurs have an aptitude and a disposition towards mental experiments and to making new combinations; they are willing to take risks, because they are not deterred by hazards that would deter the rest of us from undertaking such experiments; and they have a good understanding of how their potential customers make mental connections, and that means they have a clear picture of potential markets.

We have noted already the similarity of this perspective to the ideas of Koestler (1964) on creativity. There is also something in common here with Martin's (2007) theory that great business leaders have a rare ability to develop an 'opposable mind' and use 'integrative thinking'. This is the idea that great business strategies are born in the minds of great business leaders when they manage to hold in their minds two contradictory theories and, instead of rejecting one and keeping the other, find a resolution between the apparently contradictory theories.

Furthermore, we shall see that Shackle/Earl entrepreneurs can have a huge potential role in making the sorts of connections described in our model of innovation and wealth creation (Chapter 19). This is an area of economics which needs much more development.

Finally, note that the Shackle/Earl theory is broader in scope than the Schumpeter view. The Shackle/Earl entrepreneur may engage in all kinds of exploratory activity that would not count as innovation, and therefore could not be entrepreneurship in Schumpeter's sense.

NOTES

[1]   This section draws heavily on Earl and Wakeley (2005, Chapter 3).

11. Organisation for innovation

In this chapter we shall discuss some of the basic economic ideas about how firms need to organise themselves for innovation. We shall see that the answer to this question depends on which of two very different perspectives we take on the creative and innovative process. These are the two perspectives described in Chapter 9. We shall see that these two perspectives have very different implications: an organisation designed to facilitate the first will tend to look very different from an organisation designed to facilitate the second. In what follows we shall describe these differences in organisational design and the implications for whether large firms or small firms have an advantage in different types of innovation.

TWO PERSPECTIVES ON INNOVATION

In Chapter 9 we encountered two very different perspectives on creativity and innovation. In the first, innovation follows from specialisation and the division of labour. Once again, we remind ourselves of Smith's key idea (1776/1904a, p. 11): 'the invention of all those machines by which labour is so much facilitated and abridged seems to have been originally owing to the division of labour'. In the second, innovation follows from the combination and reorganisation of existing but previously distinct knowledge and competencies. Koestler used the term *bisociation* to describe this process of combination.

These different perspectives have different implications: first, in terms of the sorts of innovations to which they give rise; and second, in terms of the design of innovative organisations. Most of this chapter will be concerned with the implications for organisation, but let us spend a few moments on the first point.

We shall assert here that the first approach to innovation (through specialisation and division of labour) tends to produce a steady stream of *incremental* innovations while the second process (combination and bisociation) is often required for *radical* innovations. This is a bold assertion and perhaps an over-simplification. However, our justification is, in brief, as follows. In Chapter 3, we argued that the difference between incremental and radical innovation was not one of size. Rather, the issue is whether an established producer can cope comfortably with an innovation or whether that innovation is disruptive to the established producer. If the established producer can cope comfortably, then the innovation is incremental; if not, the innovation is radical.

The first approach to innovation works *within* an existing structure while the second approach requires the innovator to *cut across* existing structures. It seems likely that innovations born by working within an existing structure will on average be much less threatening to that structure than innovations born by working outside that structure. In short then, innovations of the first type (born out of the division of labour within an existing structure) are relatively likely to be incremental, while innovations of the second type (born out of combinations that cut across existing structures) are relatively likely to be radical.

ORGANISATION FOR DIVISION OF LABOUR

The first approach to innovation (specialisation and division of labour) tends to be most developed within hierarchies. This is especially true in the U-form (or functional structure) organisation. In a U-form organisation, each unit of the firm manages a particular business function (e.g. finance, marketing, or manufacturing). The motivation for such a structure is that it facilitates the division of labour, and hence allows the firm to exploit scale economies from specialisation.

The U-form structure is best suited to relatively stable conditions and stable markets with little innovation or technical change, where job descriptions are stable and most decisions are routine. This is an environment in which operational efficiency (i.e. carrying out the task at the lowest cost) is the key element of business strategy.

By contrast, the U-form structure is not well designed to cope with radical innovation, where job descriptions must change frequently and where many decisions are non-routine. Those sorts of changes conflict with accepted norms and routines, and in that business environment, U-form puts too much decision-making pressure on the CEO.

As firms grow, they often evolve from U-form into M-form – or, multidivisional form. This is a structure where the organisation is broken into units, but not along functional lines. So, for example, while the U-form organisation's units may be different functions (e.g. finance, marketing, or manufacturing) the M-form organisation's units may represent different products (e.g. mainframe computers, desktop computers, laptop computers).

For a given size of organisation, the M-form structure is, arguably, somewhat better designed than U-form to cope with innovations that call for a redefinition of routines within a specific product area. However, no hierarchy (whether U-form or M-form) is really suited to the challenges of radical innovation.[1]

ORGANISATION FOR COMBINATION

The second approach to innovation (combination and bisociation) requires a rather different form of organisation. The very idea that bisociation involves combining the *habitually unfamiliar* means that such innovation calls for communication that cuts across existing structures and communication with unfamiliar communities using unfamiliar channels. That is by definition hard to develop in hierarchies (especially U-form, but also M-form) because the required channels do not exist and to create them involves a substantial and painful organisational change. But such communication is easier to develop within a network structure.

A network structure is one in which relationships among work groups are not described by formal lines of authority but are governed by the often-changing implicit and explicit requirements of common tasks. Indeed, workers or work groups can be reconfigured and recombined as the tasks of the organisation change.

A very striking example of this was provided in a pioneering study of firms in the electronics industry. After interviewing many managers in UK electronics firms, Burns and Stalker (1961, p. 92) argued that:

> In the electronics industry ... there is often a deliberate attempt to avoid specifying individual tasks, and to forbid any dependence on management hierarchy as a structure of defined functions and authority. The head of one concern, at the beginning of the first interview, attacked the idea of the organisation chart as inapplicable in his concern and as a dangerous method of thinking about the working of industrial management.

Sometimes this network structure is created within a single company. But sometimes it is created by an alliance between several network firms. As discussed in Chapter 13, these network firms specialise in a small part of the value chain, and trade with a network of other firms to complete the vertical chain. And once again, as stressed in Chapter 13, these network firms tend to thrive in clusters or industrial districts.

The advantages of this network structure are that structures can be changed frequently and rapidly. This means that these structures are well adapted to cope with radical innovation. In addition, network firms well placed to concentrate on core competencies in house, and outsource the rest. The disadvantages of this network structure are: (a) that it encounters problems with multiple lines of authority and divided loyalties (even more than in the matrix structure); and (b) that unstable job descriptions make it hard to achieve economies of specialisation. The very factors that make this structure well designed for innovation by combination make it poorly designed to achieve innovation by specialisation.

The Essential Difference

The essential difference between these two different approaches to organisation can be summarised as follows. A hierarchical organisation has limited communication channels. Indeed, the M-form organisation is optimised to run with a minimum number of communication channels. This is quite satisfactory for incremental innovations because in creating such innovations, extensive networks are of limited value – and may even be counter-productive. By contrast, organisations that need to cope with radical innovation need organic form, flat structures and copious networking. To achieve these innovations, the company must develop a continually changing 'coupling process', and such change would put severe strains on any rigid internal structures (Freeman and Soete, 1997).

WHO HAS THE ADVANTAGE IN INNOVATION?

The simple answer to the question which forms the title of this section is, 'it depends'. It depends on what sort of innovation we are discussing. We can draw some rough generalisations about the conditions in which large companies or small companies hold the advantage in innovation.

Large companies adopt organisational structures that allow them to develop economies of scale – especially U-form (but sometimes also M-form). As Adam Smith famously observed (Smith, 1776/1904a, p. 19): 'the division of labour is limited by the extent of the market.' This means that organisations operating in large markets can benefit from specialisation and the division of labour while organisations operating in small markets have much less to gain. For that reason, large organisations, operating globally, can develop a high degree of labour specialisation, and hence are well placed to excel in incremental innovation.

By contrast, these hierarchical organisational forms are not designed to promote information-sharing across diverse functions and therefore such large organisations are not well adapted to encouraging the sort of creative combination that is needed for radical innovation. For that reason, large companies tend to find radical innovations are disruptive or competence-destroying.

When do small startups have the advantage? Startups find it easier to bring together a new combination of competencies *ab initio*. Such a recombination involves no organisational change and challenges no existing organisational structures. Startups, 'organic' firms and network firms more generally, have the flexibility that makes them well adapted to develop radical innovations.

By contrast, startups cannot achieve the same economies of scale available to large global players, or to specialised component suppliers operating on a global scale. Nor can startups achieve the same degree of labour specialisation, and are generally not well placed to challenge large players in supplying incremental innovations

THE ROLE OF VISION

Some large companies find it helpful to develop a corporate *technology vision* to help them compete in rapidly changing markets. We shall discuss the idea of vision in more detail in the Chapter 12. But in essence, it is something between a forecast and a strategic plan, stating what technological developments the company expects and what products and technologies it plans.

Why is *vision* important in helping the company organise for radical innovation? It is important because careful use of vision can help to turn what might appear to be radical innovation into incremental innovation. Why is that? Because it is easier for a company to anticipate and make plans for those innovations which it can see coming. And, as we argued above, radical innovation tends to be de-stabilising for the large hierarchical organisation while incremental innovation need not be.

So who needs a vision to help them organise for innovation? The answer to this is the same as the answer to this question: which companies find radical change the hardest to cope with? Large, mechanistic and hierarchical organisations cope with incremental change but find radical change much harder. Small, *organic* and 'flat' organisations are better at coping with radical change and therefore have less need for a vision.

A simple analogy can help to explain this point. Consider the different experiences involved in trying to navigate a narrow waterway in a large ship or a small rowing boat. In the large ship, such navigation is hazardous because the ship has so much inertia. Any change in direction has to be planned well in advance for otherwise the ship runs the risk of a collision or going aground. By contrast, in the rowing, such navigation is not hazardous because the boat has no inertia. The 'skipper' can change direction at the last moment and without much planning or risk. In this metaphor, the corporate vision is akin to a nautical chart of the waterway. The pilot of the large ship will benefit from careful inspection of the chart so that (s)he is well aware of shallow areas and can work out well in advance how to manoeuvre through the waterway without running risks. By contrast, the skipper of a 'rowing boat' has no real need for a chart.

In just the same way, the manager of the large firm (which has much inertia) will benefit much from having a vision to allow strategic planning, while the manager of the small firm (which has little inertia) can do without.

NOTES

[1] It is arguable that a matrix organisation is better adapted to radical change than either the U-form or M-form to the extent that each employee works with two lines of communication: one is to colleagues working in the same function (e.g. production) and the second is to colleagues working on the same product areas. Having said that, matrix structures bring their own management problems.

12. Vision and innovation

This chapter examines the idea of a vision of the future of a technology or a market.[1] These visions are often widely articulated within companies and sometimes widely publicised outside, especially in markets with rapid technology change. Indeed an organisation's vision could be said to be an essential part of its technology strategy.

Our concern in this chapter is less with the ex post factual accuracy of these visions; rather we shall concentrate on how the visions are articulated and publicised, and what their effects are on the subsequent development of the technology and market structure.

EXAMPLES OF VISION

Let us consider two types of vision. The first could be described as a *bold statement* that some new technology has no future. One of the best known visions of this sort is the remark attributed to Thomas J. Watson of IBM in 1943: 'I think there is a world market for maybe five computers'. Even when it become clear (later in the 1950s) that this first vision had been wrong, many mainframe manufacturers continued to assert a second vision that there would be no market for computers *other than* the large mainframe. Again that turned out to be quite wrong.

In retrospect we tend to laugh at such examples, and use them to imply a lack of foresight. One radically different interpretation is the precise opposite. Those who make these statements actually suspect that the new technology *does* have a future and it is a future in which their own business interests will suffer. So, in this case, the argument is that mainframe companies knew perfectly well that there *was* a potential market for minicomputers and ultimately microcomputers, and knew that these computers could undermine the market for mainframes. But they feared that these minicomputers and microcomputers were technologies that they could not control. Rather than give these futuristic products a level of credibility, these companies deliberately played them down. The statement that there is, 'no demand for computers other than large mainframes', can be interpreted to mean, 'there is undoubtedly a potential market for small computers, but we desperately hope it doesn't take off!'

The vision may have been wrong in the sense that it was an inaccurate *prediction* of the future. But it was the right *strategy* for a mainframe producer to play down the future of technologies that could undermine their market position and even threaten their very survival. A canny strategist should not (in public at least) admit the possibility of scenarios in which their market position is weakened.

A second example of a vision is when a company makes its future product plans public. For example, Intel Corporation (which makes the microprocessors at the heart of PCs) has frequently made use of this strategy. Well in advance of the actual product launch, the company would announce to the trade press details of their next generation of products and the projected dates on which these would be introduced.

At first sight, this seems a puzzle because these *preannouncements*, as they are called, give away vital commercial information to rivals. Why would a firm do that? The answer is that these preannouncements also have a strong influence on customers and potential customers. Suppose, for example, a customer faced a choice between using a current Intel microprocessor and a newly introduced Motorola microprocessor for some new system of their own. Suppose that this user found the Motorola product met his/her needs better. Then in the absence of any knowledge about future product introductions from Intel, the customer might choose to use the Motorola product. But if the customer had learnt from an Intel preannouncement that the latest superior Intel product would be available in a few months, then the customer might wait to use the superior Intel microprocessor when it becomes available. Indeed, if the Intel preannouncement gives away some technical information about the forthcoming product, then the customer can start to design his/her system around the forthcoming product *even if* it is not yet available.

TERMINOLOGY

It can be useful to classify visions according to two criteria. First, are they *tactical* or *strategic*? That is, are they directed to short-term objectives (tactical) or long-term (strategic)? Second, are such visions used to achieve an effect within the firm (*internal*) or outside (*external*)? The external effects could be further sub-divided into effects on different categories of *outsider*: rivals, business partners, customers, government and others.

Table 12.1 illustrates the four categories and gives examples. By short-term we mean no more than the lifetime of one generation of product; in semiconductors this may be as little as three years. By medium term we mean five or ten-year plans. Long-term visions sometimes extend to a twenty-year plan or even further ahead!

Table 12.1 Categories of corporate vision

	Tactical (e.g. preannouncements)	Strategic (long-term visions)
Internal	Commit warring factions to a particular product: 'it is in the programme ... it has been agreed'	Used in reorganising company to cope with future technology change: if vision is well understood, then change is incremental
External	User: encourage buyer to wait for new product Rival: signal, or entry deterrent	Encourage user to plan future products and production around this vision of the future of the technology

TACTICAL USES OF VISION

The tactical use of vision is typically a statement about the intention to introduce a particular product, service or production process at some point in the near future. They are often called preannouncements.

The internal and tactical use of vision is probably very widespread. Companies can use the product preannouncement to force a consensus (or at least a decision) on divided factions within the organisation. For once the CEO of a company has committed the company, in public, to a particular action then the organisation loses credibility if it does not deliver. The CEO can use this issue of company credibility to force the divided factions to settle their differences and agree to a particular product introduction.

More generally, tactical preannouncements can play a role in helping companies to establish the sorts of organisational routines essential for the operation of a hierarchical organisation. Nelson and Winter (1982) famously described the organisational routine as a *truce* in intra-organisational strife. A common example of this is the tension that exists between the marketing and design (or R&D) functions in many organisations. Marketing want a story they can sell to the market about a product to fill a niche at the right time. This can lead marketing departments to promise what design and/or R&D consider to be the undeliverable. On the other hand, design and/or R&D want enough time to develop a product with the minimal number of design faults. But at the end of this time period, marketing may consider that the market opportunity will have disappeared. The CEO can use a preannouncement to force a compromise on each party, because once the introduction date and the form of the product are public knowledge, then the company has much to lose if it fails to deliver.

External tactical use may be of at least two sorts. One is the sort of preannouncements made by Intel – as described earlier in the chapter. The company preannounces a new product to encourage buyers to wait for the better (or cheaper product) rather than buy the rival's product today. This sort of preannouncements can play an important role in the standards race – as we have already noted in Chapter 7. An interesting question arises in that context: what happens if the preannouncement is over-optimistic? What if the company announces that a product will be introduced in November 2009, but it doesn't actually appear until November 2010? This is an important issue, because preannouncements have been abused by some producers who promise their product much earlier than it actually appears. While accurate preannouncements would appear to increase the customers' knowledge, and that would appear to have positive welfare effects, it can be shown that inaccurate preannouncements can be used in a predatory (or anti-competitive) way.

The second external tactical use of vision is not directed at customers but at rivals. The company makes a preannouncement of its intentions as a signal to competitors and perhaps as a deterrent entry. A classic example of this was found in the market for computer memory chips.[2] There have been episodes in this market where firms appear to enter some sort of preannouncement *auction*. One firm announces a date and a price at which they will sell the next generation of computer memory chips. A rival may then announce an earlier date and/or a lower price. The bidding goes on until no one else is prepared to commit to an earlier date or a lower price.

The effect of this preannouncement auction can in principle be to ensure that even before any actual investments are made, the lowest bidder has effectively won the market battle. But this is only the end of the story if the winner demonstrates a credible commitment to their strategy (perhaps a large-scale investment in fabrication equipment). Without that, their preannouncement lacks credibility and entry would not be deterred.

STRATEGIC USES OF VISIONS

Thomas J. Watson's pessimistic forecast for the future of the computer and the pessimistic forecasts for mini-computers are both examples of a strategic (long-term) use of the vision. Or, in a more positive frame, some companies in biotechnology or nanotechnology have forecast very *optimistic* futures for their technology to public that may be sceptical. A strategic use of vision could take the form of the following statement: 'We expect the following broad path of developments in the technology. Product A will appear next year, product B (twice as powerful) three years later and product C (twice as powerful again) three years after that. They will observe upward compatibility.'

As with tactical uses of vision, the purpose here can be internal or external. Firms can make internal use of visions as a long-term planning device. If this vision permeates through the organisation, and is understood by all divisions in the company, then any required organisational changes along that path can be anticipated and will, if well managed, seem more incremental than radical. As argued in Chapter 11, hierarchical organisations find it easier to cope with incremental change than radical change. On the other hand, without a clearly articulated and well-understood vision, organisational change cannot be anticipated and any change is likely to appear radical. Hierarchical organisations find it much harder to cope with that. Indeed, it could reasonably be said that the distinction between incremental and radical change has little meaning unless defined with reference to the organisation's vision of the future. To some degree, visions become embedded in organisational structure, so that the extent to which organisations can cope with innovation depends on the extent to which it is anticipated.

Visions may be used for external strategic purposes, where the aim is (for example) to influence the investment decisions of users and/or the entry decisions of rivals. The difference between the external strategic and the external tactical is one of degree. Strategic use is not about getting customers to wait for *one* preannounced model. Rather, the objective is to give the customer a preview of future technology, and to encourage the customer to base their own planning and investment decisions around this long-term vision. This can be especially relevant for a technology with high switching costs, for in that setting customers may make a long-term commitment to use the technology of one particular supplier. Decisions about long-term commitments of that sort should take into account long-term projections of what each supplier can offer into the future.

While we have made these analytical distinctions between tactical and strategic, on the one hand, and between internal and external on the other, firms may in practice use visions for a variety of purposes. Indeed any one vision statement may have more than one type of effect.

NOTES

[1] This chapter draws heavily on my Chapter 3 in Swann and Gill (1993), which contains a detailed review of the literature on vision.

[2] Porter (1980, pp. 78-79).

13. Clusters and networks

This is one of the longest chapters in the book, and for good reason. First, clustering is a pervasive characteristic of innovative industries. Second, the cluster offers a powerful way to bring together many ideas from all parts of this book.

SILICON VALLEY

What is a cluster? Perhaps the easiest way to understand what a cluster entails is to look at a brief history of the leading example of a successful cluster, Santa Clara Valley in California, better known as 'Silicon Valley'.

Silicon Valley became the centre of the US (and world) computer industry in the 1950s, building on electronics expertise in Stanford University, and US military spending on electronics. It is home to many major companies in the computer industry, including Apple, Hewlett-Packard, and Intel. One distinguishing characteristic of Silicon Valley is the large number of startup companies that became major players in the industry (e.g. Intel and Apple). Many companies there are described as 'network firms', specialising in one part of the vertical chain and outsourcing the rest. It is generally reckoned to have the best risk and venture capital resources anywhere in the world. Some essential features of economic life in Silicon Valley are informal networking, job mobility, network firms, startups, specialisation and the division of labour.

The history is a very interesting one. In 1940, Santa Clara County was an agricultural valley, and even in 1950 there were no more than a few fledgling electronics companies based in the area. By the mid-1950s, however, it had become the best location for semiconductor companies to set up, and by the end of the 1960s, the area had become internationally renowned as Silicon Valley, the centre of the world semiconductor industry.

Historians of the growth of Silicon Valley identify a number of critical events and other influences in its emergence. First, World War II stimulated the research-based aircraft manufacturing industry in California. Many aircraft firms located in the state to take advantage of the good weather conditions which allowed all-year-round testing of aircraft and outdoor assembly. During the war, Federal funding was directed to Stanford University laboratories for the development of electronic components and equipment for military use.

A second critical event was the appointment of Frederick Terman as Provost and Vice-President of the university. He was an exceptionally far-sighted and entrepreneurial Vice-President, who transformed electrical engineering at this university, and did much to bring in large amounts of government and industrial funding to the university. Partly in consequence of this, many of the war-related aerospace and electronics companies which located in Santa Clara County clustered their operations around the university. One powerful examples of this was Hewlett Packard. In addition the Stanford Research Institute at the university had a mission to perform research that helped stimulate west coast business.

A third key event was the Korean War. This brought a continuing flow of funds to Stanford and firms in the area for basic electronics research and development, as well as creating a strong demand for electronics-related products. The development of ballistic missiles, for example, boosted the west coast aircraft industry and created a strong demand for components from the young semiconductor industry. A large proportion of the Pentagon's research and procurement spending came to the Stanford area – between 1950 and 1954, it is estimated that military prime contracts awarded to California accounted for about 14 per cent of all such awards nationwide.

The fourth key series of events, perhaps, occurred during the late 1940s and early 1950s, when a string of well-established electronics-related firms moved parts of their operations to the Santa Clara area. Some of the largest national firms also located R&D facilities in the county during the period. Some exceptionally influential startups were also established in Santa Clara County, including Shockley Transistor (founded by one of the three scientists who shared a Nobel Prize for inventing the transistor), and Intel (a spin-off from Fairchild) who invented and popularised the microprocessor.

Three other factors played an important role in the evolution of Silicon Valley. The first was the good weather which makes Santa Clara Valley an attractive location in which to live. Indeed it is said that one of the reasons that Terman chose to move from MIT to Stanford was because he did not enjoy the best of health, and the Palo Alto climate was much superior.

Second, Santa Clara enjoyed very good communication between engineers in the aerospace and semiconductor industries, and this was essential for the development of many essential electronic components for use in aerospace. Saxenian (1994) describes an atmosphere which was conducive to informal communication between different firms, even competitors. Many personnel in rival firms might be friends from college or university days, and would continue to meet informally to share technological knowledge. Personnel mobility between firms was high: moving jobs once a year was not uncommon. Informal communication and job mobility both helped to generate very rapid and widespread information sharing, and this was made for an environment very conducive to innovation.

Third, Santa Clara County business has enjoyed a very high quality of venture capital sources. It could be said that Santa Clara Valley enjoyed a pioneer's advantage in this respect: venture capitalists there were the first to learn about the financial side of this sort of high technology business, and better adapted to both serve and profit from it.

By the mid-1960s, military demand for electronic components was becoming less important (in relative terms) but the strong concentration of electronics companies in the area continued to attract new entry, to foster the competitive edge of firms located there, and promote rapid growth amongst incumbents. By then, Silicon Valley had established a self-sustaining momentum, and no longer needed significant government activity to sustain its growth.

As far as clusters go, this is as good as it gets. Most other clusters do not enjoy all the beneficial effects observed in Silicon Valley. Nonetheless, clustering is an essential feature of life in many industries – especially high-technology industries, so we shall discuss it at some length in this chapter.

THEORETICAL DEFINITION OF A CLUSTER

Use of the term *cluster* is not standardised, and different authors mean different things. The Table 13.1 tries to summarise in a very simple way some of the different interpretations of the cluster concept. It is a simple one-dimensional spectrum of interpretations, from rich to shallow.

Table 13.1 The cluster 'ladder': variations on the cluster concept

Phenomenon	Richness of Cluster	Difficulty of Measurement
Informal knowledge exchange	Rich	Difficult
Explicit collaboration	↑	↑
Labour mobility		
Marshallian externalities		
Network firms		
Companies interdependent in a value chain		
Co-location and superior performance		
Co-location and technological proximity		
Co-location	Shallow	Easy

Start at the bottom of Table 13.1. The most 'shallow' definition of clustering simply says that a group of firms are co-located. A slightly more demanding definition is that this co-located group should also be technologically related (e.g. they are in the same industrial sector – as in Silicon Valley). The next step upwards would require that these co-located firms should show superior performance and that this is attributable to their location in a cluster. A further step would be that these firms in the cluster are explicitly inter-related in a value chain, as for example in Figure 13.1.

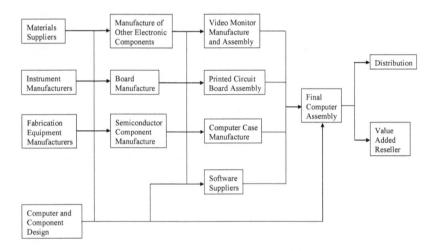

Figure 13.1 Simplified value chain for personal computer manufacture

So for example, if the cluster contains all the firms illustrated in Figure 13.1, and they trade directly with each other as illustrated, then that would be a further step up the ladder of Table 13.1. The next step up the ladder is that firms in the clusters have an explicit strategy as 'network firms' and exploit the existence of all the other network firms in the cluster. Network firms are firms that specialise in a very narrow part of the vertical chain, and outsource most other activities. Such network firms are common in strong industrial clusters and benefit from location in strong clusters (networks) when: (a) the required competencies are uncommon and no one team member has them all; and (b) the ability of team members to work with each other cannot be taken for granted, so it helps to have a large pool to draw on. We discuss the idea of network firms in more detail below.

The next step up is that firms in the cluster enjoy the sorts of mutual benefits from each other's company that are usually called 'Marshallian externalities' (e.g. sharing a common infrastructure, access to a pool of skilled labour, and so on). A further step up would be that there is labour mobility around this network, and this is important because mobile labour is one of the most effective ways of transferring technology around all the firms in a cluster. A further step up would be that companies in the cluster have explicit collaboration in R&D – not, we should stress, in anything that might get them into trouble with anti-trust authorities. And the final step of Table 13.1 is the sort of informal knowledge exchange between technologists of different, even *rival* companies in the bar after work.

Some qualitative researchers, in particular, say that we only have a cluster when we have almost everything on the list in Table 13.1. From their point of view, the items at the bottom of table may be *necessary* conditions for a cluster to exist, but they are not *sufficient*. By contrast, some quantitative researchers tend to interpret something as a cluster even if it does not climb so far up the ladder in Table 13.1. This difference is *directly* related to the relative ease of measuring the phenomena at the bottom of the ladder and the relative difficulty of measuring phenomena at the top of the ladder. Quantitative researchers (such as econometricians) tend to focus their attention on phenomena that can be measured quantitatively. This is the case for the phenomena at the bottom of the ladder but is rarely possible for the phenomena at the top. For the most part, quantitative researchers cannot say how near a cluster gets to the top of the ladder, because the last few rungs cannot easily be measured. By contrast, qualitative researchers (such as those who write case studies or carry out in-depth interviews with company executives) will aim to focus on all phenomena on the ladder, and are better placed to say exactly how far a cluster has progressed up the ladder.

EXAMPLES OF CLUSTERS IN UK

No cluster in the UK enjoys the sheer cluster energy of Silicon Valley. In fact, that remark would apply to any other cluster in the world. Indeed, some have suggested that even Silicon Valley itself does not have the same cluster energy in the computer industry as it had in the past. Be that as it may, there are some important clusters in the UK and here we give a brief description of seven leading examples.

Probably the strongest cluster in the UK is the financial services cluster in the City of London. About half a million people are employed in financial services in the City of London, and London accounts for more than 50 per cent of UK employment in the management of financial markets, venture capital, security broking and trading, fund management and bank HQs. The Corporation of London Report (Taylor et al., 2003) recognises many benefits of a City of London location in financial services, including: the importance of London as a credible business address; proximity to customers, skilled labour and professional bodies; access to knowledge; and the wider attraction of London as a major city. Four factors ensure the continuing growth and vitality of the financial cluster in London: good labour supply; good personal relationships through face-to-face contact; support for innovation; and the advantages of co-location and competition. However, it is recognised that there are also some disadvantages of a London location, notably the high cost of premises and the shortcomings of the public transport system. These factors are leading to some 'de-clustering' of back-office, routine administrative procedures to locations outside the City of London, but companies still find it essential to keep a City office.

A large share of the UK TV, media and music industries is located in a cluster within London. London has 55 per cent of UK employment in TV, located in Westminster and Hammersmith, much of it around the BBC. It has an even greater proportion of the UK film industry (over 70 per cent) and the music publishing industry (75 per cent). Indeed, these clusters are interdependent with other creative industry clusters in London (advertising, publishing, clothing and fashion). Total creative industry employment within London is estimated at half a million (the same as the financial services in the City of London. The report on *Creativity: London's Core Business* (GLA Economics, 2002) recognises that despite their diversity, these creative industries function in very similar ways: they have high growth rates of output and employment, they have a strong tendency to cluster in London and these industries depend on a very strong supply of a particular type of human input – creative and intellectual labour.

Much of the information and communication technology (ICT) industry is clustered in what is called the 'M4 Corridor' after one of the major motorways in the south of England.[1] The industry, like the motorway, runs from West London to South Wales, but the most important parts of this 'corridor', as far as the ICT industries are concerned, are between West London and Swindon (Wiltshire). This includes important parts of the electronics sector, computer industry, telecommunications, software and digital content.[2] Broadly defined, the 'M4 Corridor' accounts for nearly half the UK's ICT industry and indeed, some people talk of the M4 Corridor as 'the UK's Silicon Valley', but this is a considerable exaggeration. Even if we add up the entire UK ICT industry we do not get such a strong industry as is to be found in Silicon Valley itself.

The UK biotechnology industry shows relatively strong clustering in what is often called the 'Golden Triangle', the area between London, Cambridge, and Oxford. A report by the Minister of Science identified the following factors as essential for the success of biotech clusters:[3] a strong science base;

entrepreneurial culture; growing company base; ability to attract key staff; availability of finance; premises and infrastructure; business support services and large companies in related industries; skilled workforce; effective networks; and a supportive policy environment.

The ceramics industry cluster around Stoke-on-Trent employs some 23,000 people. Pottery accounts for about three quarters of that figure. This cluster dominates UK production of pottery and ceramics. It accounts for 80 per cent of total UK employment in household and ornamental goods, 60 per cent of total UK employment in technical products and insulating ceramics, and 40 per cent of UK employment in the production of ceramic tiles, flags and sanitary fixtures. Although this is a long established and mature industry, Staffordshire University still plays an important role in strengthening the cluster with programmes in ceramic design and research on the technical testing of ceramics. [4]

The North West chemical cluster in Cheshire and Merseyside employs about 44,000 people.[5] This cluster is the largest chemical cluster in the UK, accounting for about a quarter of UK employment in chemicals and is the North West region's most important single industry. There has been a chemical industry in the region since Roman times, based on local salt deposits. The industry in the NW cluster is much more diverse than in other areas of the UK (Grangemouth, Teesside, or Humberside) which are based around a few major multi-national companies. The NW cluster has a high proportion (about 80 per cent) of small and medium enterprises (SMEs) and these smaller companies depend on the surrounding cluster.

Our final example is of clustering in the brewing industry around the town of Burton upon Trent (in the West Midlands). For centuries, Burton upon Trent was a preferred centre for brewing beer because of the quality of water available, which contains a high proportion of dissolved salts. Originally, English drinkers favoured darker beers (stout and porter) which were not easy to transport and hence the market in any town was supplied by local breweries. But the development of rail links and the change in taste to bitter led to a concentration of the industry around Burton.[6] Bitter was easier to store and transport, and the larger breweries, by exploiting economies of scale and falling transport costs, could out-compete smaller local breweries. These factors and a series of mergers and takeovers led to a concentration of three main breweries and, at its peak, Burton produced a quarter of all beer sold in Britain. In the early 1970s, there was something of a reaction to mass-produced beers with the emergence of CAMRA,[7] and a resurgent taste for 'real ale' meant that the industry became slightly more dispersed again.

WHY DO COMPANIES CLUSTER?

The company's location decision is, like many other economic decisions, one where there is some element of choice at an initial stage, but by the time the company is firmly established in a particular location it is much less likely to move – even if there would be some benefits in doing so. So we need to answer the question in two parts. Why do companies making location decisions tend to cluster? And why do well-established companies tend to stay where they are? The first is the more interesting one. The second is often a case of a lock-in. It would be too expensive to relocate, so that any benefits from relocation are outweighed by switching costs.

Table 13.2 Advantages and disadvantages of clustering

	Demand Side	Supply Side
Advantages	• Strong local customers • Reduced customer search costs • Market share gains from clustering (Hotelling) • Reduced transaction costs • Information externalities	• Strong local suppliers • Pool of specialised labour and other specialised inputs • Shared infrastructure • Reduced transaction costs • Information externalities and knowledge spillovers • Facilitates innovation
Disadvantages	• Competition in output markets	• Competition in input markets (real estate, labour) – 'overheating' • Local infrastructure over-stretched • Congestion (e.g. in transportation) • Cartels • 'New ideas need new space'

Source: Adapted from Swann et al. (1998, p. 57), Taylor et al. (2003).

Companies making location decisions cluster for a variety of reasons. We can group the benefits from clustering into two types: demand side and supply side. We should also note that while there are, indubitably, advantages from clustering, there can also be disadvantages, and these can also be broken into demand side and supply side disadvantages. Table 13.1 summarises these advantages and disadvantages from the perspective of the clustered firm.

Advantages on the Demand Side

Some companies benefit from having strong local customers for their products and services. At first, this may seem surprising. After all, in an age of low transport costs and online services, many companies supply global markets and their customers are spread all over the world. The existence and strength of local customers might not appear to matter very much.

However, this takes too simplistic a view of the relationship between an innovative company and its customers. As we shall see in Chapter 15, we can make a distinction between *passive* consumers and *active* consumers. The consumers of conventional mainstream economics are reasonably passive. The producer understands their needs and any innovation undertaken to supply these needs is undertaken by the producer. The consumer just goes to the market and buys. However, many consumers for innovative products and services are not like that. They are much more active. As von Hippel (2005) put it, innovation is democratic: customers are involved in innovation as well as producers.

Companies do not, on the whole, need close geographical proximity with their *passive* consumers. They can ship their products round the world to these consumers and provide online after-sales support. But companies can benefit enormously from close contact with their *active* consumers, because active consumers are often an important source of ideas for the next generation of innovative products. Some companies find considerable competitive advantage in co-locating with some of their key customers – or at least having *an office* near key customers – even if other parts of the operation are located elsewhere.

A related idea is that companies may benefit from location in a cluster because that reduces the search costs of potential customers. Again, in the age of the Internet, that might seem irrelevant. If all suppliers list their entire catalogue online then the customer can search and find what (s)he is looking for regardless of the location of the supplier. However, that assumes that the products and services can be adequately described online and that in turn assumes a reasonable degree of standardisation. If I seek a particular model of digital (DAB) radio, then I can find it online, and probably even order online. But if I seek a very particular antique, which cannot be fully described online, so that I must examine it carefully before purchase, then it is most efficient to look for antiques in those areas of major cities where antique shops cluster together, or at the (temporary) clusters formed at the international antique fairs. Location in a cluster means that the discerning consumer will be more likely to search my store to see if (s)he can find what (s)he wants. Location outside the cluster means that the customer is much less likely to come across my store accidentally, and as a result is much less likely to find what I have in store.

A third demand-side benefit from clustering is the idea captured in Harald Hotelling's (1929) famous old model of the two ice-cream sellers located on a beach. This model examines the location decisions of two ice-cream sellers on a beach of 1 km in length where holiday-makers are distributed uniformly along the length of the beach. The model assumes that there is no price competition or product differentiation: both sellers have the same product range and set the same prices for their ice creams. The model also assumes that demand is inelastic: each customer will, if necessary, walk the full length of the beach to reach the nearest ice-cream seller to buy what (s)he wants. The model assumes that customers will for convenience always choose the nearest seller. Finally, the model assumes that relocation is costless: either seller can wheel his/her barrow to a new position on the beach if (s)he thinks there is competitive advantage in doing so.

Under these conditions, the model shows that the equilibrium outcome is for the two sellers to cluster together side by side at the mid-point of the beach. If they were located apart, then that would not be an equilibrium, because either producer would stand to gain market share by bringing his/her barrow closer to his/her rival, and thereby gaining some more custom from holiday-makers located in between his/her barrow and his/her rivals. If they are located together, but not at the mid-point of the beach, then one seller will have less custom because (s)he only supplies less than half the beach. In that case, the seller would have an incentive to move his/her barrow to the other side of his/her rival, so that (s)he is now the preferred seller for more than half the beach. But any such advantage is at most transitory, because the rival will repeat the same move. This progressive leap-frogging will go on until they are clustered together at the centre. That is the only point at which each seller is in an equilibrium position

A fourth demand side benefit is the existence of information externalities. This is the idea that if I see another trader selling successfully at a particular location then that tells me something about the strength of local demand. The other trader's visible success creates an information externality. This seems to be a strategy employed by some café proprietors. They see that a particular café in a particular location is performing well and reckon that if an adjacent property becomes available, this would be a good location in which to open a rival café. In this example, there is an element of location to reduce consumer search costs. If the new café is located next to a well-known and popular established café, then the newcomer will be easier for customers to find, and may benefit if the established café is very busy.

Finally, clustering can reduce transaction costs more generally – not just the reduction in search costs described above. The literature on the economics of organisation argues that transaction costs may be important when: (a) it is a difficult task to ensure that components from an external supplier will exactly meet the customer's requirements; b) it is costly to communicate with outside companies; c) the customer is concerned about the risk of opportunistic behaviour by sub-contractors.

Each of these concerns and costs may be reduced when both parties to the transaction are located in a cluster. The argument is slightly different in each case. (a) If the input required is complex and it is difficult to ensure that inputs are compatible with the company's requirements, then the outsourcing company will need a lot of face-to-face contact with the sub-contractor. This is much easier and cheaper in a cluster, because the outsourcing company and its sub-contractor are physically close to each other. (b) More generally, if it is difficult to communicate requirements with sub-contractors, then it helps to build up familiarity with the sub-contractor. Communication between those who know each other well is generally more effective than between those who do not. It is easier to form lasting business relationships in a cluster. (c) Sub-contractors within a strong cluster may have a disincentive to behave opportunistically because they know that they may earn a bad reputation within the cluster. They also know that any opportunistic behaviour may mean that they lose repeat business with important customers in the cluster. For that reason, the risks of opportunistic behaviour (and hence the hold-up problem) may be lower in the cluster.

Advantages on the Supply Side

Now we can turn to the advantages to clustering that derive from the supply side. The first benefit is the existence of strong local suppliers. Again, as with the first type of demand-side benefit, we might wonder if this really matters all that much. As we argued before, in an age of low transport costs and online services, many companies buy their inputs from global markets and their suppliers are spread all over the world. The existence and strength of local suppliers might not appear to matter very much. But as we argued above in the context of local customers, this is too simplistic a view. In those cases where the company is a passive customer of standardised components, then it can obtain these components from anywhere. But when the company is an active customer for non-standardised components, then it may require regular face-to-face contact with its suppliers and that is easier if they are co-located in a cluster.

The second, and long-recognised, supply side advantage to location in a cluster is that it means the company has access to a large common pool of specialised labour and other specialised inputs. If the company has a very specialised requirement, then it will be more likely to find this in a large cluster than in an outpost of the industry. The reason is simply that a specialised worker, for whose services demand is limited, will generally find it most efficient to locate in a cluster because that is where the jobs will be. We can say, indeed, that this is both a demand-side and a supply-side advantage to clustering.

The third benefit is similar, and relates to the fact that clustered companies can share a common infrastructure which is not available to companies outside the cluster. This shared infrastructure could be very wide in scope, including transport infrastructure (road, rail, and air), public assets (the science base and other publicly provided business services) and real estate (suitable office buildings). In surveys, quite a lot of companies say that this logistic argument is an important reason for location in a cluster, even if they do not enjoy any of the other benefits from clustering.

The fourth benefit is reduced transaction costs. Transactions with a neighbouring supplier may be easier than transactions with a distant supplier. The argument here is just the same as that above – in the context of demand-side benefits.

The fifth benefit is information externalities and knowledge spillovers. These spillovers can operate on the supply side in the same way as they operate on the demand side, as discussed above. Companies may learn from informal knowledge exchange with their neighbouring suppliers just as they learn from informal knowledge exchange with their neighbouring customers.

Finally, proximity with suppliers can facilitate the innovation process. It is well known from surveys and case studies of innovation that customer-supplier interaction plays an important role in the innovation process. Moreover, in the *combinatorial theory of creativity* (described in Chapter 9), creativity requires the inventor to bring together habitually distinct insights. While this need not involve social networking between distinct people, it often does. To the extent that such creativity involves bringing together exactly the right mix of people with distinct but complementary expertise, then such creative work is easier to achieve in a cluster and network with a wide diversity of participants. To facilitate such common understanding is generally reckoned to be very difficult online but may be easier when the two groups meet face to face on a regular basis. This is easier to achieve in a cluster than at a distance. The case study below gives an interesting illustration of how clustering enhances innovation.

Disadvantages on the Demand Side

The above paragraphs indicate that there is much to be gained from clustering. However, as indicated in Table 13.1, clustering can bring some disadvantages to the clustered firm as well as advantages.

The first of these, which may be relevant in some cases, is that the clustered firm may encounter greater competition in the local markets it supplies than it would if it were located outside a cluster. Obviously this argument is most relevant when the customers that matter are mainly local. (If a clustered company supplies a global market, then greater competition in supplying local customers is probably not of great concern to it.) Using a common English metaphor, we can say that the clustered firm is a small fish in a large pond while the non-clustered form is a large fish in a small pond. If we apply a simple Cournot model to analyse this question, we find that the equilibrium price is an inverse function of the number of competitors. Customers benefit through lower prices when there is lots of competition in a cluster, but suppliers suffer because the prices they can charge are lower.

Disadvantages on the Supply Side

Most of the disadvantages of clustering apply to the supply side, however. The first is that firms located in a cluster may face more competition in their input markets. The most obvious examples of this are the greater competition for real estate and for skilled labour in a cluster. This may sometimes lead to what is informally called 'overheating'. Thus for example, financial services companies located in the City of London face very high rents for their offices. This is certainly a disadvantage to locating in that cluster, and is a factor in the de-clustering of some parts of the value chain. Another example, in the context of Silicon Valley, was the rapid growth in the salaries that skilled semiconductor engineers could command. This is again a disadvantage to locating in that cluster, and once again was a factor explaining the relocation of some parts of the computer industry value chain to other parts of the USA, or indeed to other (lower wage) economies.

A second, and related point, is that in a strong cluster the local infrastructure shows signs of being over-stretched. The City of London again provides a striking example of this. Parts of the London Underground date to the nineteenth century and were not built to deal with the volume of passengers that now use that system. A third aspect of this is that clusters become uncomfortably congested. This is a problem that many living in London, Tokyo and other major world cities complain about. Moreover, the road infrastructure around major clusters tends to become seriously congested as the cluster grows – the South East of England and Los Angeles are two examples.

A fourth disadvantage of location in a cluster is that the clustered firm may suffer from the existence of supply-side cartels. Just as geographical proximity makes it easier for companies to collaborate in research and innovation, so it makes it easier for companies or other agencies to collude in their supply of a critical input. Collaboration of the first sort is usually legal, and indeed encouraged by government. Collusion of the second sort is often illegal, but such collusion still goes on, and the severity of anti-trust legislation varies from country to country. The following are a couple of examples of such collusion. First, if the land around a cluster is owned by a small number of landowners they may collude to keep down the supply of land for development and thereby keep up real estate prices and rents. This is reckoned to be an issue in some high-tech clusters. Second, trade unions in clusters may have acquired more power over working hours and conditions, and the company locating in a cluster may find that the available labour, while skilled, is less flexible than the labour available elsewhere.[8]

A fifth and final disadvantage of location in a cluster is captured by the maxim: 'new ideas need new space'. This idea has its roots in the *autonomous theory of creativity* – see Chapter 9. To some degree at least, creativity means breaking the rules and those who do that will usually encounter resistance from their peers. As peer-group contact is much more vigorous in a cluster than outside a cluster it may be easier to break the rules in isolation. Indeed, the autonomous theory of creativity says that deep creativity requires that inventors either have (or develop) a degree of social and/or emotional autonomy which means that either they are isolated from such peer group pressure or they just ignore it. A leading example of how the development of a new product innovation required new space can be found in the development of the IBM PC – which has become the PC standard in widespread use today – see the case study below.

CASE STUDY 1: CLUSTERING ENHANCES INNOVATION

This case study refers to a British company that designs and supplies state-of-the-art electronic components. The company designs its components in the UK but fabrication is done in South East Asia. The company has been first to market with many of its product lines – and this is seen as essential to its competitive strategy.

The company's management believes that it is simply not an option to try to achieve cost competitiveness in 'low end' components. The 'low end' components are commodities: they are all standardised, many companies can make them, and the customer does not mind much where the components come from. Price competition is intense. In this company's view, the only possible area of competition for a UK-based component producer is in the 'high end' components that the 'low end' producers are not able to produce. To produce these 'high end' components requires an active programme of innovation. Moreover, in a matter of months, the 'low end' producers will be able to copy the 'high end' products of yesterday, so that is why continuing innovation is so essential, and indeed why it is so important to be first to market. These innovations are often innovations in the product but can also be innovations in marketing. The most important innovations are those that involve articulating the next generation of products.

One of the company's key ideas in understanding the essential role of innovation and the importance of being first to market is a 'continuous loop model' of innovation. The company that is first to market with one generation of chip will enjoy an advantage in production of the next chip. This is because that company will have the first customers to experience the component and can therefore have the first dialogue with customers about their experiences. This in turn gives the company a head start in envisaging the next generation of products, and hence a head start in the race for the next product innovation. Since a key part of the company's strategy depends on early articulation of the next generation of chips, and since that requires regular and in-depth dialogue with key customers, then it is important to have a location next to the key customers. As many of the key customers for this company are located in the USA, it was important for the company to establish a local base in the USA. It is interesting to see that a traditional reason for clustering – to do with the necessity of co-locating design and fabrication – is not very important for this company.

CASE STUDY 2: INNOVATION OUTSIDE THE CLUSTER

IBM was not the first pioneer in the commercial market for computers. However, when the first UNIVAC computers appeared in 1951, IBM was quick to quick to realise the commercial potential, and by 1956, IBM had taken an 85 per cent share (by value) of new systems sold. Although with the advent of minicomputers and plug-compatible computers, IBM's market share fell back from 85 per cent, IBM still accounted for a half of the entire US computer industry's revenues in the late 1970s. IBM was important not just in the final market for complete computers, but was also highly vertically integrated, making most of the components, peripherals and software in its computer systems.

But in the mid-1970s, the personal computer arose as a severe challenge to IBM's market dominance. To begin with, IBM did not take it seriously, but by 1979, senior management at IBM realised that they must rise to the challenge of the personal computer and produce their own version, and quickly. The IBM employee charged with working out their strategy for entering the PC market, William Lowe, realised that the only way they could do this quickly was to work around the IBM bureaucracy. If they had built the PC according to the standard IBM formula, where each part was produced by an IBM division, the results would have been much too slow. The only way to do this quickly was to procure components, peripherals and software from outside suppliers. Although this strategy represented a radical challenge to the IBM way of doing business, senior management endorsed it and the code name, 'Project Chess' was given to this PC project.

The project team was located at IBM's Boca Raton labs in Florida, to put them at a safe distance from many of the principal manufacturing divisions of IBM. The PC team was allowed to act as a startup company, with complete autonomy, and with IBM as a venture capitalist. To meet the short deadline required not only bypassing the bureaucracy, but also breaking the IBM convention of internal sourcing. Not only was the microprocessor used (the Intel 8088) a non-IBM component, but parts of the actual fabrication of the PC were done by outside suppliers. When IBM divisions complained about this, they were told that they could submit bids like anyone else, and indeed some divisions did win contracts. But most of the major components of the system were bought in from outside suppliers. The two central parts of the IBM system, the 8088 microprocessor (from Intel) and the operating system (from Microsoft) were bought in, and the producers of these enjoyed huge benefits from the success of the IBM PC – benefits, some would argue, that could have been kept within IBM.[9]

NETWORK FIRMS

In the discussion above, we have talked about the high incidence of 'network firms' in clusters. They are firms that take advantage of the efficiency gains available though the division of labour to specialise in a very narrow part of the vertical chain, and outsource most other activities. Such network firms are common in strong industrial clusters, as they benefit especially from location in strong clusters.

The reader of this book might be thinking: can the theory of networks (described in Chapter 7) be applied to clusters? The answer is yes, up to a point, but we need to be clear that clusters and networks are not necessarily the same. To see this, consider Figure 13.2.

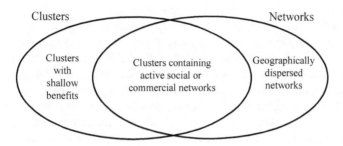

Figure 13.2 Clusters and networks

The aim of this is to show that the concepts of network and cluster overlap, but are not identical. Some 'rich' clusters contain active commercial networks, and some networks have geographical proximity; both of these are illustrated by the intersection of these two sets. But some 'clusters' do not enjoy active social networks: these are clusters (according to the shallow definition of Table 13.1) but are not networks. Equally, some networks are not geographically concentrated: these are networks without being clusters.

With that qualification, it is useful to revisit the three network 'laws' defined in Chapter 7: Sarnoff's Law, Metcalfe's Law and Reed's Law. In brief, these three laws stated:

- Sarnoff's Law (broadcast networks): aggregate value of a network is proportional to number of members (n)
- Metcalfe's Law: (communication networks): aggregate value of a network is proportional to the square of number of members (n^2)
- Reed's Law: (group forming networks): aggregate value of network is an exponential function of number of members (2^n).

We shall argue that Metcalfe's Law often applies to clusters and in some creative industry clusters, Reed's Law may apply – up to a point.

The literature on economics of organisation uses the term 'network' in two senses: the network structure and the network firm. A *network structure* within a firm is one in which relationships among work groups are governed more by the often-changing implicit and explicit requirements of common tasks than by formal lines of authority. Indeed, workers or work groups can be reconfigured and recombined as the tasks of the organisation change. A *network firm* is one which specialises in a small part of the vertical chain, and trades with a network of other firms to complete the vertical chain. The network structure into which network firms locate themselves (as in the previous paragraph) will change as the activities of firms alter. A classic example of the network firm was Apple in the early days of the PC market (late 1970s onwards). Apple was a highly specialised design company, which outsourced most component supply and indeed much of the assembly process to other network firms in the Silicon Valley. The structure of the network which Apple used changed as their competencies and strategy changed.

Network firms thrive in clusters because there they can find a wide variety of possible partner firms and sub-contractors from which to form the network required to complete the vertical chain of production. Moreover, the prevalence of face-to-face contact in clusters, coupled with the prospects of repeat business for honest traders, may mean that firms in a cluster enjoy relatively low transaction costs. Moreover, network firms are well placed to exploit efficiency gains from the division of labour, concentrate on core competencies in house, and outsource the rest. So the network firm is well suited to markets with rapid innovation.

This organisational form also has some shortcomings. Hybrid forms such as network firms, strategic alliances, and joint ventures, require careful coordination. But the loose evolving structure of these hybrid forms may compromise coordination. There may not be clear formal management structures, nor mechanisms to make decisions and resolve disputes. These forms depend on a high degree of trust and reciprocity, and can suffer from agency costs (e.g. risks of *free riding*) and influence costs.

Kay (1997, p. 205) said of the joint venture that it 'is an extremely expensive device through which to coordinate resource allocation, and ... it is typically only resorted to as a device of last resort'. Such hybrid forms are used when the cost of assembling an equivalent collection of competencies in house is prohibitive, and when the difficulties of using 'arms length' market transactions are too great. These hybrid forms are used when:

- Technological change is very rapid
- Market opportunities are transitory or of uncertain longevity, so long-term contracts or merger are unattractive
- No one party has expertise to do everything in house, or it would be excessively costly to organise this internally
- All parties have to make relationship-specific investments, so there can be potential hold-up problems
- Transactions are hard to pin down with comprehensive contracts
- Transactions are complex, not routine. Traders cannot count on contract law to 'fill the gaps'.

The network firm has a particular advantage when it is part of a large clustered network from which a wide variety of specialised teams can be assembled. In this case, Metcalfe's Law or even Reed's Law applies to the network firm. That is, the value of network membership increases rapidly with the size of the network.

The network firm in a large clustered network can bring together a wide range of competencies to fill a short-lived market opportunity. To repeat what we said before, this is especially relevant where:

- the required competencies are uncommon and no one team member has them all
- the ability of team members to work with each other cannot be taken for granted, so it helps to have a large pool to draw on.

THE CLUSTER LIFE CYCLE

The advantages and disadvantages from clustering, as summarised above, tend to occur at different stages of the history of the cluster. Many of the advantages tend to occur during the early history of the cluster and many of the disadvantages tend to occur during the later history of the cluster. These observations give rise to the idea that clusters exhibit something like the life cycle observed with products.

Statistical and econometric evidence suggests that there are some important sources of positive feedback during the early history of a cluster. This evidence suggests that:

- Companies located in strong clusters often grow faster than average
- Strong clusters attract disproportionate amounts of new firm entry (startups)
- In high-tech industries (e.g. biotechnology), proximity of the science base (e.g. a major university) attracts entry
- Strong clusters generate high levels of innovation and patenting.

This evidence suggests a pattern of positive feedback as described in Figure 13.3. Clusters with a strong industrial base and a strong science base attract entry and promote growth. That entry and growth in turn strengthen the industrial base and science base in the cluster. That in turn gives a further boost to entry and growth – as shown. And so on.

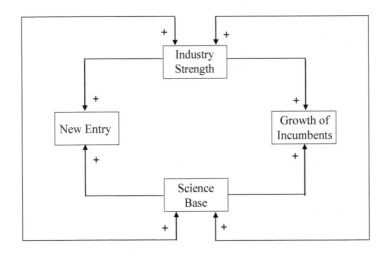

Figure 13.3 Positive feedback in the growth of clusters

But these effects tend to get weaker as the cluster gets older. Moreover, the econometric evidence suggests that the effects of cluster strength on productivity and financial performance are weaker. In summary, clustering seems *more* important for those activities that are important in the introductory and growth stages of the product life cycle (such as entry, growth, invention, innovation) while clustering seems *less* important for those activities that are important in the maturity and decline stages of the product life cycle (such as productivity and cost-cutting).

How big does a cluster have to be to enjoy benefits? Some writers have used the concept of 'critical mass', borrowed from physics, to describe this. One way to get a handle on critical mass is by looking at rates of new firm entry in different cluster sizes. Swann et al. (1998) showed that such entry patterns in the US computer industry followed a life cycle curve looking roughly like that in Figure 13.4.

In small clusters, entry is modest. As the cluster grows, however, the rate of entry increases, and at an accelerating rate. Beyond a certain point ($E'' = 0$), the rate of entry continues to grow, but the rate of acceleration is tailing off. And eventually, we reach a point of maximum entry ($E' = 0$). Beyond that point, there is still entry into the cluster, but at a declining rate.

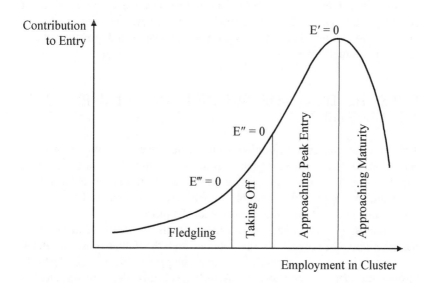

Figure 13.4 Critical mass in clusters

This life cycle curve marks three points on the curve:[10]

- The maximum rate of acceleration in entry ($E''' = 0$)
- Zero acceleration in entry ($E'' = 0$)
- The maximum rate of entry ($E' = 0$).

This curve suggests that there is a concept of critical mass in cluster size, and that the benefits of clustering do not accrue until the cluster reaches a certain size. It also suggests that these benefits tail off as the cluster gets large. This means that there is a limit to the positive feedback illustrated in Figure 13.3.

To the right of the peak of the curve in Figure 13.4, the cluster is approaching maturity. This means that the disadvantages of clustering are starting to catch up with the advantages of clustering and will eventually overtake the advantages. We encounter 'overheating' and congestion in clusters, as described above. Real estate prices and rents may become impossibly expensive and labour costs may be very high. Local infrastructure may start to be over-stretched and inhabitants of the cluster suffer from the pervasive congestion. We find that while entrants may continue to benefit from entry, their entry does not convey any benefits to incumbents within the cluster and may simply impose further congestion costs on incumbents. Mature clusters may become less attractive places to locate because of the rise of certain forms of cartel behaviour and because the environment is not conducive to more radical innovation. As all these things happen, the feedback from growth in cluster size to entry, innovation and growth becomes zero and eventually negative.

DOES THE 'DEATH OF DISTANCE' MEAN THE END OF THE CLUSTER?

The last topic concerns a question about clusters that is the source of much confusion. Does the advent of the Internet, low-cost global communications and low transport costs mean that the concept of cluster is becoming irrelevant? Much journalism and other contributions to popular debate would suggest that it is. But such arguments are too simplistic.

Cairncross (1997) coined the memorable term 'the death of distance' to describe how the advent of the Internet and the falling cost of global communications and transport means that companies find it possible to do business at ever greater distances. At one level, this 'death of distance' is undeniable. But, surprising as it may seem, that does not necessarily mean that location becomes less important for economic activity. Nor does it imply the end of clustering.

A simple way to understand this is to consider Figure 13.5 on the following page. This illustrates a model of the geographical evolution of an industry as transport costs decline. In particular, the model illustrates why the decline in transportation costs may lead to *greater* geographical concentration of production rather than a dispersion of production.

It applies to a context where economies of scale are important and where there is little or no product differentiation between the products produced in different locations. So for example, it can be applied to developments in the brewing industry up to 1975 where (as we saw above) the greater transportability of 'keg' ales, the limited product differentiation and important economies of scale led to greater concentration in the brewing industry with a major cluster in Burton upon Trent in the English Midlands (see above). The model assumes that consumer demand for beer (in aggregate) is price inelastic but relative demand for different brands of beer *is* quite price elastic. The model does not apply so well to the brewing industry of today where discerning consumers have a well-developed demand for 'real ale' and product variety, which means that product differentiation is much more important than in this simple model, and the trend towards geographical concentration is not so strong.

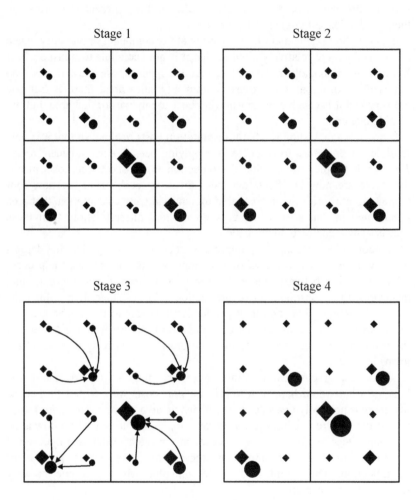

Figure 13.5 The 'brewery model' of agglomeration

Figure 13.5 represents the first four stages in the evolution of the industry. Each part of the diagram represents a geographical map showing the distribution of population (the diamonds represent different towns and cities) and the distribution of beer production (the circles represent breweries and their production volume) at different times.

Stage 1 shows the pattern of production at the beginning of our story, when transport costs are relatively high, so that it is not economic to transport beer from one town to another. To capture that, we superimpose a grid on the map representing different sales areas. Within any sales area, there is but one brewery and it is capable of transporting beer to any part of that area, but not outside that area.

Now in Stage 2, suppose that transportation costs decline, so that it is now feasible for breweries to compete in supplying beer to some adjacent areas. To keep it simple, assume that the reduction in transportation costs means that there are now in effect four sales areas. In each, there are now four competing breweries that are capable of supplying any customers in that area – but *only* in that area. If there is no product differentiation, then these breweries will compete on price alone.

If economies of scale are important, then it is likely that the largest brewery in each area will have an advantage over the others, and the process of competition between these breweries will lead (after mergers and takeovers) to a consolidation of production. This is Stage 3, in which the smaller breweries close and their production migrates to the largest brewery in the area. The outcome of that consolidation is shown in Stage 4. Each of the (now larger) sales areas has just one brewery, supplying beer to all the towns in the area.

While the graphs just illustrate these first four stages, we can easily imagine what would happen next. If transport costs were to fall further, so that there is now only one sales area (the whole area shown in the graph), then any of the four remaining breweries can supply beer to any of the towns on the map. Once again, in the absence of any product differentiation, and if economies of scale are important, there will be a further process of consolidation. This could lead to the outcome where just one brewery supplies the whole map.

Obviously, this simple model glosses over a number of factors. In particular, if there is product differentiation, and discerning customers demand variety, then several breweries could survive even if one becomes the largest. But the model shows the basic idea quite clearly. Reduction in transport costs does not lead to *greater dispersion* but to *greater concentration* of production.

This outcome may puzzle some readers. Surely the reduction of transport costs makes it easier for the small town brewery to compete in the big city?

Yes, indeed, so it does. But it also makes it easier for the big city brewery to compete in the small town. The reduction in transportation costs removes a source of competitive distinction between the different breweries. Location *as such* is no longer an important source of competitive distinction in the eyes of the consumer. But location now determines economies of scale, costs and prices, and prices *are* important in the eyes of the consumer.

So we come to this slightly paradoxical conclusion: falling transport costs makes location less important from one perspective, but more important from another. This sort of paradox can be found in many analyses of how falling costs of transport and communication influence the geographical location of activity. The same sort of result occurs whenever:

1) Location as such of the producer is unimportant to the customer.
2) Falling costs of transport and communication increase the extent of the market and hence increase (in the short-term, at least) the number of competitors in the market.
3) In an increasingly competitive global market, the survival of a specific company depends ever more on that company exploiting every possible source of competitive distinction.
4) If there is any source of competitive advantage in location then the company should exploit it.
5) Companies in clusters can often find sources of competitive advantage that are denied to companies located outside clusters.

In these conditions, location is less important from one point of view but more important from another point of view. Moreover, in these conditions, falling costs of transport and communication can make clustering *more* (rather than *less*) important. Indeed, production may become even more concentrated into a smaller number of clusters, and companies located in the periphery will decline.

We do not suggest that falling transport and communications costs will always lead to this paradoxical result. If points (a) to (c) above apply but points (d) and (e) do *not*, then firms in the periphery may perform just as well as those in clusters, and dispersion of economic activity is viable. But if (d) and (e) apply, then we should not expect reduction in costs of transportation and communications to reduce clustering and disperse activity. To put it another way, falling costs of transportation and communication with no reduction in economies of agglomeration will lead to *greater* clustering. The dispersion of economic activity requires decline in *agglomeration and scale economies* – not just a decline in the costs of transport and communication. But in many cases these agglomeration and scale economies are, if anything, getting stronger!

NOTES

[1] http://www.seeda.co.uk/
[2] http://www.sqw.co.uk/
[3] http://www.berr.gov.uk/sectors/biotech/
[4] http://www.thepotteries.org/
[5] http://www.chemicalsnw.org.uk/
[6] http://en.wikipedia.org/wiki/Burton_upon_Trent_brewing
[7] http://www.camra.org.uk/
[8] This is said to be one of the reasons why the Japanese car makers locating plants in the USA and UK chose to locate their plants well away from the traditional centres of car manufacture. In the USA, Japanese companies chose not to set up plants in Detroit but preferred less densely populated areas with less of a union tradition (Rubinstein, 1992). In the UK, Nissan chose not to locate in any of the traditional car making areas (East London, West Midlands and Merseyside), but chose instead to locate in the North East.
[9] Langlois (1992) gives a more detailed account of the story of the IBM PC.
[10] These are, respectively, the points at which the third, second and first derivatives of the entry function are set at zero.

14. Division of labour

We have already met the concept of the division of labour at various points in the book. In Chapter 2 we noted Adam Smith's observation that the division of labour can give rise to invention and this was a theme we revisited when discussing theories of creativity and invention in Chapter 9. In Chapter 11, we observed that those organisational forms designed to exploit potential efficiency gains from the division of labour would be best suited to generate the sorts of incremental innovations that stem from the division of labour. And in Chapter 13, we saw that network firms in clusters could take advantage of the efficiency gains from the division of labour to specialise in a narrow part of the value chain.

In this chapter we shall take an in-depth look at the relationships between the division of labour and innovation. We stress the plural, *relationships*, because the relationship is bi-directional. First, the division of labour is one of the determinants of invention. But second, innovation in turn leads to a division of labour. In what follows we shall make much reference to the work of Adam Smith, John Rae and Charles Babbage and this is evidence of what we asserted at the beginning of the book: economists always have something to learn from re-reading the works of the greats in the history of economic thought.

ADAM SMITH ON THE DIVISION OF LABOUR

While he was not the first to discuss the economic and social role of the division of labour, Adam Smith was probably the first to put the division of labour at the centre of a discussion of economic growth. Indeed, the division of labour gets pride of place in his 'Wealth of Nations' – probably the most influential book ever written in economics. I quote him at length (Smith, 1776/1904a, pp. 5-6):

> The greatest improvement in the productive powers of labour, and the greater part of the skill, dexterity, and judgment with which it is anywhere directed, or applied, seem to have been the effects of the division of labour ... To take an example, therefore, from a very trifling manufacture; but one in which the division of labour has been very often taken notice of, the trade of the pin-maker; a workman not

educated to this business (which the division of labour has rendered a distinct trade), nor acquainted with the use of the machinery employed in it (to the invention of which the same division of labour has probably given occasion), could scarce, perhaps, with his utmost industry, make one pin in a day, and certainly could not make twenty. But in the way in which this business is now carried on, not only the whole work is a peculiar trade, but it is divided into a number of branches, of which the greater part are likewise peculiar trades. One man draws out the wire, another straights it, a third cuts it, a fourth points it, a fifth grinds it at the top for receiving the head; to make the head requires two or three distinct operations; to put it on is a peculiar business, to whiten the pins is another; it is even a trade by itself to put them into the paper; and the important business of making a pin is, in this manner, divided into about eighteen distinct operations, which, in some manufactories, are all performed by distinct hands, though in others the same man will sometimes perform two or three of them. I have seen a small manufactory of this kind where ten men only were employed, and where some of them consequently performed two or three distinct operations. But though they were very poor, and therefore but indifferently accommodated with the necessary machinery, they could, when they exerted themselves, make among them about twelve pounds of pins in a day. There are in a pound upwards of four thousand pins of a middling size. Those ten persons, therefore, could make among them upwards of forty-eight thousand pins in a day. Each person, therefore, making a tenth part of forty-eight thousand pins, might be considered as making four thousand eight hundred pins in a day. But if they had all wrought separately and independently, and without any of them having been educated to this peculiar business, they certainly could not each of them have made twenty, perhaps not one pin in a day; that is, certainly, not the two hundred and fortieth, perhaps not the four thousand eight hundredth part of what they are at present capable of performing, in consequence of a proper division and combination of their different operations.

In short, the division of labour plays an absolutely central role in the growth of productivity and hence in wealth creation.

Lest we think there is anything special about the example that Smith chose, it is interesting to compare it with the following extract describing the same phenomenon in watch-making, written about thirty years earlier (in 1747):[1]

At the first appearance of watches they were but rude to what they are now; they were begun and ended by one man, who was called a watch-maker. But of late years the watch-maker, properly so called, scarce makes any thing belonging to a watch; he only employs the different tradesmen among whom the art is divided, and puts the several pieces of the movement together, and adjusts and finishes it.

The next improvement watches and clocks received, was the invention of engines for cutting the teeth in several parts of the movement, which were formerly cut by hand. This has reduced the expense of workmanship and time to a trifle, in comparison to what it was before, and brought the work to such an exactness that no hand can imitate it.

The movement-maker forges his wheels of brass to the just dimensions; sends them to the cutter, and has them cut at a trifling expense: he has nothing to do when he takes them from the cutter but to finish them and turn the corners of the teeth. The pinions made of steel are drawn at the mill, so that the watch-maker has only to file down the pivots, and fix them to their proper wheels. The springs are made by a tradesman who does nothing else, and the chains by another: these last are frequently made by women, in the country about London, and sold to the watch-maker by the dozen for a very small price. It requires no great ingenuity to learn to make watch-chains, the instruments made for that use renders the work quite easy, which to the eye would appear very difficult. There are workmen who make nothing else but the caps and studs for watches, and silver-smiths who only make cases, and workmen who cut the dial-plates, or enamel them, which is of late become much the fashion.

When the watch-maker has got home all the movements of the watch, and the other different parts of which it consists, he gives the whole to a finisher, who puts the whole machine together, having first had the brass-wheels gilded by the gilder, and adjusts it to proper time. The watch-maker puts his name upon the plate, and is esteemed the maker, though he has not made in his shop the smallest wheel belonging to it.

Watch-making in the mid-eighteenth century experienced a division of labour to match what we find in modern manufacture of the personal computer – see below.

ADAM SMITH ON INVENTION

One of the important consequences of the division of labour, as Smith saw it, was that it gave rise to invention and innovation. (Smith didn't make a distinction between invention and innovation – as we have done above.) For Smith, the key thing here is that invention stems from a prior division of labour. That is, the direction of causation runs from the division of labour to the invention. Here we give the full quote from Smith – we have already quoted an abbreviated version (Smith, 1776/1904a, p. 11):

> The invention of all those machines by which labour is so much facilitated and abridged, seems to have been originally owing to the division of labour. Men are much more likely to discover easier and readier methods of attaining any object, when the whole attention of their minds is directed towards that single object, than when it is dissipated among a great variety of things. But in consequence of the division of labour, the whole of every man's attention comes naturally to be directed towards some one very simple object. It is naturally to be expected, therefore, that some one or other of those who are employed in each particular branch of labour should soon find out easier and readier methods of performing their own particular work, wherever the nature of it admits of such improvement. A great part of the machines made use of in those manufactures in which labour is most subdivided, were originally the inventions of common workmen, who, being each of them employed in some very simple operation, naturally turned their thoughts towards finding out easier and readier methods of performing it.

Smith sees the direction of causation here from division of labour to invention and innovation. From this perspective, invention and innovation are not activities that call for much networking. Specialised labour builds up enough experience through learning by doing from which to create inventions as a problem-solving exercise. It is interesting to compare this with the *autonomy* theory of creativity described in Chapter 9.

Marx also saw a direct connection from the division of labour to the direction of technical change (Marx, 1867/1974, p. 323):

> so soon as the different operations of a labour-process are disconnected the one from the other, and each fractional operation acquires in the hands of the detail labourer a suitable and peculiar form, alterations become necessary in the implements that previously served more than one purpose. The direction taken by this change is determined by the difficulties experienced in consequence of the unchanged form of the implement Manufacture is characterised by the differentiation of the instruments of labour – a differentiation whereby implements of a given sort acquire fixed shapes, adapted to each particular application, and by the specialisation of those instruments, giving to each special implement its full play only in the hands of a specific detail labourer ... The manufacturing period simplifies, improves, and multiplies the implements of labour, by adapting them to the exclusively special function of each detail labourer. It thus creates at the same time one of the material conditions for the existence of machinery.

Thus the process of the division of labour creates an incentive for specialised labour to seek to modify their tools and invent new ones. The rate and direction of invention is influenced by the extent to which existing tools are unsuited to specialised use. Hence, for Marx, the division of labour shapes technical progress, but equally he recognised that developments in the division of labour followed prior technological change.

JOHN RAE: THE REVERSE DIRECTION OF CAUSATION

For Smith, invention was an important consequence of the division of labour, but it is the latter that lies at the heart of economic growth and wealth creation. As we saw in Chapter 2, John Rae (1834) was perhaps the first economist to put invention at the heart of economic growth. And for Rae, the direction of causation was from invention to the division of division of labour, and not vice versa. Referring to Smith's 'Wealth of Nations', he put his alternative perspective as follows (Rae, 1834, Appendix to Book 2):

> In the Wealth of Nations, the division of labour is considered the great generator of invention and improvement, and so of the accumulation of capital. In the view I have given it is represented as proceeding from the antecedent progress of invention.

Rae doubted that the division of labour could give rise to invention – for reasons that will become clear below. But why did he think that the direction of causation should run from invention to the division of labour? Rae did not have Smith's clarity of expression and for that reason his message has not had the same impact on the history of economics. But it is an important message.

His explanation centres around economies of scale and specialisation. Suppose an invention creates a new piece of capital equipment, which can offer a substantial productivity gain. But suppose this piece of capital is expensive and it is only really economically viable for a firm to buy one if it is in regular use. Rae then compares two forms of industrial structure: one in which each workman does a little of each activity, so there is no division of labour. The other in which each workman specialises in a particular operation, so there is a well developed division of labour. Rae points out that in the first case it will not be viable for each (or indeed for *any*) workman to buy his own piece of equipment so the invention will be unused. But in the second case it will be viable for the one specialist workman to buy the equipment and put it to constant use. In that second case, the capital equipment can be produced and used with consequent gains in productivity and wealth creation.

So Rae's argument is in effect that productivity-enhancing inventions will only be adopted and used if a specialist workman puts the invention to constant use. And as these productivity-enhancing inventions can be wealth creating, then an entrepreneur will see the profit opportunity to be made from a division of labour.

Who is right? Smith or Rae? Not for the first time in this book, I shall hedge my bets and say that both of them can be right. Instead of demanding that one is right and the other wrong, I think we should recognise that there is something in both arguments. Certainly, if we look at the development of invention, there is a role for the division of labour in creating such invention. This is very evident in the history of academic invention, if I may call it that. Equally if we look at this history of labour specialisation, we shall see investment in certain expensive machinery (or human capital) is only viable if it can be put to constant professional use. A doctor will only find it viable to invest in learning about a specific and narrowly defined illness if there is a professional opportunity to practise in that specialism alone, which probably means working in a large teaching hospital rather than as a general practitioner. An academic economist will only invest in a new research area if (s)he perceives the opportunity to profess such a specialism. If we are obliged to teach all aspects of economics (as in a one person department, for example), then it is not viable to be too specialised.

Indeed, this idea is very similar to Smith's idea that the division of labour is limited by the extent of the market. So, as many have said, perhaps the Rae view is not incompatible with Smith, even if the Smith view is contradicted by Rae. This is not, however, the same as saying that Smith was wrong and Rae was right.

THE BABBAGE PRINCIPLE

Charles Babbage is best known as one of the pioneers of computing, having invented some of the first calculating machines. In his quest to raise money to develop his invention, he visited many manufacturers in order to get a better understanding of their processes and in the hope that they might provide financial support for his invention. Unfortunately for him, they did not. But fortunately for economics, a by-product of these visits was a very remarkable book on industrial economics. In that, he articulated what is now known as the 'Babbage Principle' (1835, pp. 175-176):

> the master manufacturer, by dividing the work to be executed into different processes, each requiring different degrees of skill or of force, can purchase exactly that precise quantity of both which is necessary for each process; whereas, if the whole work were executed by one workman, that person must possess sufficient skill to perform the most difficult, and sufficient strength to execute the most laborious, of the operations into which the art is divided.

This is somewhat different from Smith's arguments. Smith says that the division of labour increases productivity because of what we would now call 'learning by doing'. If someone specialises in a narrowly defined task they get better and better at doing it quickly and accurately. But Babbage asks us to suppose a particular production process involves two tasks: one requires great skill but little strength, while the other requires great strength but no skill. Without any division of labour, it is necessary to find a worker who has both skill and strength. But with a division of labour, the strong worker can do the task requiring strength while the skilled worker can do the task requiring skill.

The Babbage effect makes it easier to assemble a world-class cricket team, for example. If all players had to be world-class 'all rounders' (i.e. world class at batting and bowling) then it would be difficult to find enough world-class players to make up a world-class team. But in practice this is not necessary, and the team is usually made up of five world-class batsmen, three to four world-class bowlers, one or two all-rounders and a specialist wicket-keeper.

DIVISION OF LABOUR IN PC MANUFACTURE

The computer industry provides a very powerful example of the international division of labour at work. Consider Table 14.1 which lists the origins of the various components of a computer system sold in the UK.

Table 14.1 Origin of components for typical PC system

Component	Origin
Brand	USA
Assembly of main box	Ireland
Chips on motherboard	USA, Korea, Taiwan, Philippines
Battery	Philippines
CD-ROM drive	China
CD-R (consumables)	Germany
Hard disk drive	Singapore
3.5″ disk drive	Philippines
Modem card	Netherlands
Graphics card	China
Specialist video card	USA
Monitor	UK
Keyboard	Mexico
Mouse	Mexico
Child's mouse	Taiwan
Loudspeakers	Malaysia
Microphone	Mexico
Inkjet printer	Spain
Laser printer	China
Zip drive	Malaysia
Scanner	Taiwan
Webcam	China
Power supplies	Taiwan, China, Malaysia, Mexico
Manuals	Scotland, Ireland, Wales, Germany
Network switch	China

Source: Author's own investigations (2004).[2]

It is almost meaningless to ask in what country a PC is manufactured.

Some have suggested that if you open up a PC and make a list of the origins of the components, the list reads a bit like the United Nations. That is something of an exaggeration, of course. Notably, in a European PC, we find no components from South America or from Africa. But as the table above shows, we still find a large number of countries are represented within a PC.

The various components are manufactured in many different countries, assembly may be done in more than one country, and the final 'badge' may be added somewhere else again. In fact this table may understate the extent of the international division of labour. Some of the components inside the PC may be labelled with the name of one country (on the outside) but inside there are sub-components from other countries, which we do not necessarily see. Particular companies in particular countries may become specialised in just one particular part of the overall process.

It was not always like this. In the early 1960s, when IBM dominated the computer industry, computer companies tended to be vertically integrated. Indeed, IBM had a high degree of vertical integration, and made almost all the components of its computers 'in house'. This included the semiconductor components, peripherals (disk drives, tapes etc.), software, operating systems, and indeed computer assembly. This started to change in the early 1970s with the advent of 'plug compatible' computers and peripherals, but the real wave of vertical disintegration started with the personal computer industry. For example, Apple Computer, founded in Silicon Valley (see Chapter 13) was one of the influential pioneers in the PC market, and at the beginning was little more than a design company, specialising in computer design. Apple produced no components and did almost no assembly (apart from the famous example of their first computer which Jobs and Wozniak built in a garage). Apple used standard components and outsourced assembly to other companies. Indeed, Apple was once described as the ultimate network firm (see Chapter 13).

In the early history of the PC market, most of the division of labour was within Silicon Valley. It is often argued that transaction costs are lower within a cluster and this means that outsourcing to specialist suppliers in the same cluster would be economically efficient while outsourcing to specialist suppliers in other countries would not.

Why should transaction costs be lower within clusters? It is helpful at this stage to remind ourselves of some of the main facets of transaction cost. While it might make sense for companies to outsource production of some input if an outside specialist could produce it better or cheaper, these companies may not do this if transaction costs are large. Amongst the transaction costs that may be important are the following: (a) complexity of coordinating inputs with company requirements; (b) costs of communicating with outside companies; and (c) risk of opportunistic behaviour by sub-contractors. It seems likely that each of these can be reduced within a cluster. The argument is slightly different in each case.

If the input required is complex and it is difficult to ensure that inputs are compatible with the company's requirements, then the outsourcing company will need a lot of face-to-face contact with the sub-contractor. This is much easier and cheaper in a cluster, because the outsourcing company and its sub-contractor are physically close to each other.

More generally, if it is difficult to communicate requirements with sub-contractors, then it helps to build up familiarity with the sub-contractor. Communication between those who know each other well is generally more effective than those who do not. It is easier to form lasting business relationships in a cluster.

Sub-contractors within a strong cluster may have a disincentive to behave opportunistically because they know that they may earn a bad reputation within the cluster. They also know that any opportunistic behaviour may mean that they lose repeat business with important customers in the cluster. For that reason, the risks of opportunistic behaviour (and hence the hold-up problem) may be lower in the cluster. For these and other reasons, transaction costs may be lower in a cluster.

Today we observe a highly international division of labour. This internationalisation has been driven by a very high degree of standardisation in the industry. The existence of these standards reduces transaction costs over distance. It does this because the ability to specify a standard in a contract reduces the risk from coordination problems – one of the key reasons for transaction costs. This, coupled with the decline in costs of transport and communication, has lead to the outsourcing of ever more component production and assembly to lower wage regions and economies. In a global market, any one company tends to specialise in the production of a small number of components to maximise economies of scale. We observe a lot of intra-industry trade between different clusters – and a lot of long-distance transport and communications!

Once again, we find that Adam Smith anticipated such developments. As noted before, Smith (1776/1904a, p. 19) argued that, 'the division of labour is limited by the extent of the market'. Students sometimes have difficulty in understanding what he meant by that, but this industry offers a perfect example. If you can supply a global market, then it can make sense to specialise in one very narrowly defined component and make it in very large volume so enjoying economies of scale. If you can only supply a small local market, then such a strategy would not make sense: the market is not big enough.

THE DAMAGING EFFECTS OF DIVISION OF LABOUR

We conclude this chapter with a rather different perspective on the division of labour. The discussion by Smith and Babbage sees the division of labour as a powerful force for economic efficiency and growth. But some critics were dubious. Yes, the division of labour might increase the productivity of labour, narrowly defined, but at what cost in terms of the loss of human dignity? One of the most vocal critics was John Ruskin (1904/1996a, p. 196):

> We have much studied and much perfected, of late, the great civilised invention of the division of labour; only we give it a false name. It is not, truly speaking, the labour that is divided; but the men:—Divided into mere segments of men—broken into small fragments and crumbs of life; so that all the little piece of intelligence that is left in a man is not enough to make a pin, or a nail, but exhausts itself in making the point of a pin or the head of a nail.

Ruskin was not the first to criticise the division of labour in this way. Indeed, Adam Smith – though, as we saw above, much impressed with the power of the division of labour as an engine of economic growth (Wealth of Nations, Book I) – had some harsh words to say on the subject in Book V of the Wealth of Nations (Smith, 1776/1904b, p. 267):

> In the progress of the division of labour, the employment of the far greater part of those who live by labour, that is, of the great body of the people, comes to be confined to a few very simple operations, frequently to one or two. But the understandings of the greater part of men are necessarily formed by their ordinary employments. The man whose whole life is spent in performing a few simple operations, of which the effects are perhaps always the same, or very nearly the same, has no occasion to exert his understanding or to exercise his invention in finding out expedients for removing difficulties which never occur. He naturally loses, therefore, the habit of such exertion, and generally becomes as stupid and ignorant as it is possible for a human creature to become. The torpor of his mind renders him not only incapable of relishing or bearing a part in any rational conversation, but of conceiving any generous, noble, or tender sentiment, and consequently of forming any just judgement concerning many even of the ordinary duties of private life.

Later, political economists and neoclassical economists were also clear about this. Jevons (1878, pp. 41-42) recognised that:

> There are certainly some evils which arise out of the great division of labour now existing in civilised countries ... In the first place, division of labour tends to make a man's power narrow and restricted; he does one kind of work so constantly, that he has no time to learn and practise other kinds of work. A man becomes, as it has been said, worth only the tenth part of a pin; that is, there are men who know only how to make, for instance, the head of a pin.

Some optimists have said that these damaging effects only apply to mind-numbing manual labour. However the sociologist Durkheim (1893/1984, p. 307), in a famous book on the division of labour, argued that these damaging effects would apply equally to a mathematician spending too much time working on a very narrow type of equation, as to a pin-maker making one tenth of a pin:

> If we have often rightly deplored on the material plane the fact of the worker exclusively occupied throughout his life in making knife handles or pinheads, a healthy philosophy must not, all in all, cause us to regret any the less on the intellectual plane the exclusive and continual use of the brain to resolve a few equations or classify a few insects: the moral effect, in both cases, is unfortunately very similar.

In view of that, we can now understand why Rae argued that the direction of causation ran from invention to the division of labour – and not, as Smith said, from division of labour. For if the division of labour is so damaging to the intellect, how can divided labour sustain any creativity or invention? And that is indeed a puzzle in Smith's argument. Historians of thought have offered some resolutions of the puzzle, but that lies beyond the scope of this chapter.

NOTES

[1] This extract is from Campbell (1747/1969). A more accessible copy of this extract is reprinted in Symonds (1947, pp. 56-57). I have modernised some of the punctuation and spelling to make this 1747 prose easier to read.

[2] If we were to repeat this investigation today, the proportion of components from China would have increased.

PART IV

Innovation and the consumer

15. The passive consumer and the active consumer

This chapter and the next are concerned with what the theory of consumption and demand has to say about innovation. Consumption and demand are huge topics in their own right and could not be given an exhaustive treatment in this book.[1] Rather, our aim is more limited: we focus just on what the theory of demand and consumption has to say *about innovation*.

This is one of the chapters of the book where we draw on a wider literature than economics alone. Why? Because the standard neoclassical analysis of demand and consumption is too limited in scope for our present purposes. In saying this we do not mean that the economic theory of the consumer is *wrong* as such. Rather, we just mean that to understand fully the way in which consumers react to innovations, we need to recognise that some consumers behave in a rather different way from that suggested by conventional economic theory.

Indeed, we shall meet *six* broad theories of consumer behaviour in this chapter, and there are important differences between each pair of these. On reading this the student may wish to sigh and say: 'Six! But which is right and which are wrong?' However, the reader should resist the temptation to seek one right theory amongst the six. None of them are wrong as such: rather, each theory applies to a different episode in consumption and demand behaviour. In some cases, we may behave according to standard economic theory, in some cases we may behave according to Galbraith's theory, and in some cases we may behave according to Veblen's theory.

We shall find it helpful to talk of six different consumers: the economic consumer, the routine consumer, the Galbraith consumer, the Douglas consumer, the Marshall consumer and the Veblen consumer. Each consumer behaves according to a different theory of consumer behaviour. Now it is quite possible that each of us will, at different times and in different circumstances, behave like each of these different consumers. But by identifying these six different types, and describing the different attitudes of each to innovation, it is easier to see how each of these plays a different role in developing a demand for innovations.

PASSIVE AND ACTIVE CONSUMERS

One important distinction is worth making at the start, between *passive consumers* and *active consumers*. The distinction is especially important in understanding the demand for innovation.

The distinction is a difficult one to capture in a few words. So let us start by illustrating the distinction with an example. Consider the different newspapers available in a newsagent. We see a spectrum from the fairly 'low brow' tabloid newspapers to the rather 'high brow' weekly or monthly publications (examples in the UK would include the *New Statesman* and the *Spectator*). The former are designed to be relatively undemanding to read and contain lots of photographs, skilful use of headlines and short, punchy news stories. The latter, by contrast, can be very demanding to read. They consist primarily of text, few photographs and often contain some quite complex argument.

It is possible for a passive consumer to read the tabloid newspaper. That sort of reading is not hard work; it does not call for any great *activity* and can almost be an automatic or passive process. The passive consumer will get something out of reading the tabloid without putting in much activity. By contrast, it is not possible for a passive consumer to read the 'high brow' weekly. To read this calls for a great deal more thought and hard work. The value that an active consumer gets from such reading depends in large measure on the amount of effort (s)he puts in.

However, whether a particular consumer is active or not is not defined exclusively by reference to the amount of activity (s)he puts in during actual consumption. A consumer may be active in seeking out exactly the right purchase but less active when consuming that purchase. More generally, at each stage in the process of demanding, purchasing and consuming, the consumer could be passive or active.

As we look at the six theories of consumption below, and the sorts of consumer they describe, we shall see that we cannot unambiguously classify these consumers as *passive* or *active*. The economic consumer is *active* in optimising his/her purchasing behaviour but may be quite *passive* in actual consumption. By contrast, the Veblen consumer is very active in actual consumption but may be fairly passive in purchasing. On the other hand, in some cases the classification is clearer: the routine consumer is passive in most things while the Marshall consumer is active in most things. But even though these different consumer types may be part active and part passive that does not diminish the value of the distinction. The distinction between active and passive consumers remains important because these two types respond to innovation in rather different ways.

ECONOMIC CONSUMER

The first of our six theories of consumption is the basic neoclassical theory of consumer behaviour. We shall call a consumer who behaves like this an *economic consumer*.

This neoclassical theory of demand and consumption is based on utility theory. The basic idea is as follows. The consumer is assumed to have a complete preference ordering over all possible bundles of goods and services and this preference ordering can be summarised by the function $u(x)$. In this neoclassical theory of consumption, the consumer is an optimiser. Choice is constrained maximisation, and while the constraints always bite, the consumer nevertheless has a large degree of discretion. The applied economics of consumption depends therefore on the assumption of revealed preference: we make inferences about preferences by observing consumption behaviour.

There are two difficulties here, however. First, does choice really reflect preferences alone? Many sociologists would dispute whether observed behaviour does actually reveal much about preferences. This difference in perspectives is beautifully summarised by Duesenberry: 'economics is all about choices, while sociology is about why people have no choices.'[2] The usual economist's response to this is to observe that, certainly, choices are constrained, but the economic theory of consumer choice can be adapted to embody more and more subtle constraints if need be. All we have to assume is that the consumer has some, even if not very much, discretion. That seems a reasonable assumption in many circumstances.

Second, to use revealed preference, it is necessary to assume that the preference *function* to be revealed is the same for all the data used in its estimation. But if these data come from different years, or represent the aggregate behaviour of an aggregate whose composition changes from one data point to the next, then this is a strong assumption. Moreover it has had the unfortunate effect (unfortunate, at least, in the opinion of this author) of focusing attention on fixed consumer tastes and away from the reasons why tastes may change. Indeed, it was only really in the 1950s and 1960s, notably with the work of Becker and others, that the endogeneity of tastes came back onto the mainstream economic agenda – and then, essentially only in theoretical work, and much less so in empirical work.

The consumer of mainstream economic theory, as described here is an unexciting individual. (S)he is an isolated optimiser: an asocial hermit of fixed and pre-determined tastes. His/her behaviour is not, apparently, influenced by others. (S)he has no need to experiment, but given the same products, prices and income would continue to consume in the same way indefinitely. In that sense, (s)he has little need for variety, though the standard assumption of convexity in consumer theory will tend to mean that the consumer consumes a collection of different goods, and does not just consume one good to the exclusion of all others.

While the economic consumer is very *active* in making the optimum decision of what to buy and how much of it, (s)he appears to be pretty *passive* in actual consumption. In short, the economic consumer is probably not very exciting company.

What does the economic consumer make of innovation? (S)he will welcome cost-reducing process innovations because these allow him/her to buy ever more goods and services from his/her income. (S)he will welcome product innovations that offer him/her more of product characteristics that (s)he values. But (s)he will not be interested in product innovations that add new characteristics which (s)he never needed.

The economic consumer may be dull but (s)he is certainly no fool. (S)he knows what (s)he wants and cannot be coaxed into thinking that (s)he needs anything else. As such, the economic consumer is resistant to the more persuasive forms of advertising. We shall now meet a consumer who is not so resistant.

GALBRAITH CONSUMER

We shall call our second consumer the Galbraith consumer after the great Canadian-born economist, J.K. Galbraith. While he has been one of the most influential economists in the wider world (holding major US government positions during the administration of J.F. Kennedy), his work has not always been met with enthusiasm by his professional colleagues. Nevertheless, his theory of the consumer is so important it must be included here.

In his famous book, *The Affluent Society*, Galbraith (1958) cast doubt on whether most consumers were as resistant to persuasive advertising as the economic consumer. Instead, Galbraith described a mass-market buyer whose wants and tastes were shaped by advertising. As Galbraith pointed out, the moment we recognise that wants are shaped by the activity of advertisers then we have to be more sceptical about the merits of economic growth (Galbraith, 1958, pp. 152-153):

> As a society becomes increasingly affluent, wants are increasingly created by the process by which they are satisfied. This may operate passively. Increases in consumption, the counterpart of increases in production, act by suggestion or emulation to create wants. Or producers may proceed actively to create wants through advertising and salesmanship. Wants thus come to depend on output.

Galbraith calls this a *dependence effect,* and notes that if production creates the wants it seeks to satisfy then it is unclear that welfare is higher at a greater level of production and consumption.

Many people would probably want deny that they ever conform to this type of consumption. And yet, it is hard to see why the advertising industry would have grown to its present size if advertising were *never* effective in shaping wants.

At this point it is useful to remind ourselves of an essential distinction in the literature on advertising between *informative* and *persuasive* advertising. The former seeks to inform the consumer: for example, the information could be that a product or service exists, that it has been redesigned and improved, or that the price has been cut. The latter seeks to persuade the consumer that they want the product or service in question, or seeks to transform the consumer's perception of the good or service.[3] The economic consumer does not ignore informative advertising: it can help him/her optimise his/her purchasing behaviour. But (s)he is unmoved by persuasive advertising. By contrast, the Galbraith consumer is heavily influenced by persuasive advertising.

While the economic consumer is *active* in buying behaviour, the Galbraith consumer is *passive* and impressionable. When it comes to actual consumption, Galbraith's consumer can be active or passive.

What does the Galbraith consumer make of innovation? The answer would appear to be that because this consumer is impressionable and malleable, (s)he could be encouraged to buy many new products and services which the hard-nosed economic consumer might not consider.

DOUGLAS CONSUMER

The last consumer was one whose tastes and wants were not fixed but could be shaped by advertisers. Now we meet a third type of consumer who, once again, has tastes that are not fixed, but the forces that shape his/her tastes are different.

The economic anthropologist Mary Douglas has been one of the most influential writers on consumption. Her writing stressed that much consumption is an inherently social activity, and yet much of the economic analysis of consumption treats it as a matter of individual choice, neglecting interactions between consumers.

Douglas wrote that 'the real moment of choosing is ... choice of comrades and their way of life'.[4] Once that choice is made, choices over lesser matters are largely determined by group norms. This is an especially interesting perspective because it suggests some *ex ante* discretion in consumption, but much less *ex post* discretion. The Douglas consumer has discretion when deciding which tribe to join but less choice thereafter. From this perspective, consumption of goods is a way of associating with 'comrades' (or peers).

Two tongue-in-cheek examples illustrate the extent to which consumption is not necessarily a reflection of choice but a reflection of powerful social norms. First, we might expect that such a powerful figure as the British Prime Minister would have the sovereignty of choice over what (s)he wears, but not necessarily. It is said that Margaret Thatcher thought that the colour red suited her best, but as leader of the Conservative party (*strictly blue*) she did not have many opportunities to wear red!

On the other side of the political spectrum, Tony Benn was in political terms *strictly red*. But on one famous occasion he was seen sporting a *blue* rosette. Why? This was on the occasion when Chesterfield FC played in the FA Cup Semi-Final in 1997. As local MP for Chesterfield, Benn attended the match and wore a rosette in the colours of his team: *all blue*! This is not so much an example of group norms over-riding personal preference but one group norm over-riding another. At an FA Cup Semi-Final, club colours are far more important than political colours!

Because the loyal Douglas consumer has limited discretion over what (s)he buys, (s)he may seem to be relatively *passive* as far as purchasing decisions are concerned. But because the Douglas consumer shares consumption as a social activity with members of his/her peer group, then the Douglas consumer may be relatively *active* in consumption itself.

How does a Douglas consumer react to innovation? At first sight it would appear that the Douglas consumer must be rather resistant to innovation, but that is not necessarily so. Innovations may offer the Douglas consumer an opportunity to affirm membership of his/her peer group in an imaginative way. Moreover, when an opinion leader within a group adopts an innovation, it is likely that Douglas consumers will be quick to follow.

VEBLEN CONSUMER

An individual's consumption can be influenced by the behaviour of at least three groups: a peer group, with which the consumer wishes to associate, and with which the consumer wishes to share consumption activities; a distinction group (or groups) from which the consumer wishes to distinguish him/herself; and an aspiration group, to which the consumer aspires to belong, but membership of which is as yet unattainable. For the Douglas consumer, the key reference group is the peer group, though some of the consumption behaviour of some Douglas consumers may also be driven by a desire to distinguish themselves from rival groups.

Now we shall turn to a fourth type of consumer whose consumption is driven primarily by this desire for distinction. For simplicity, we shall call this a Veblen consumer after Veblen's (1899) theory of conspicuous consumption: this consumer wishes to signal his/her distinction and wealth by a very conspicuous display of consumption. But as we shall see, several writers have recognised the same phenomenon so we could equally well name this consumer after them.

While Adam Smith was indeed the founder of modern economics, his consumers are capable of greater flamboyance than the traditional economic consumer. Indeed, Smith was well aware of some of the interdependencies in demand, and to some degree anticipated Veblen's (1899) concept of conspicuous consumption:[5]

> With the greater part of rich people, the chief enjoyment of riches consists in the parade of riches, which in their eye is never so complete as when they appear to possess those decisive marks of opulence which nobody can possess but themselves

Nassau Senior (1863), an influential classical economist, and first holder of the Drummond Chair of Political Economy at Oxford recognised two important features of the consumer. First, he observed:[6] 'Strong as is the desire for variety, it is weak compared with the demand for distinction, a feeling which ... may be pronounced to be the most powerful of human passions.' Senior also drew attention to:[7] 'the desire to build, to ornament and to furnish – tastes which, where they exist, are absolutely insatiable and seem to increase with every improvement in civilisation.'

The idea of distinction as a driving force in consumption has been most thoroughly developed in the modern sociological analysis of consumption by Bourdieu (1984). As Bourdieu shows, and this is not a point brought out in Veblen, distinctive consumption does not necessarily have to be very expensive consumption. It just has to be distinctively different. Thus Bourdieu shows how those of sophisticated musical taste demonstrate that sophistication by conspicuously listening to sophisticated music. (Bourdieu cites J.S. Bach's 48 Preludes and Fugues for *Well Tempered Clavier* as an example of 'sophisticated' music.) These people of sophisticated taste would, equally, go out of their way to shun low-brow taste in music. (Bourdieu cites Johann Strauss's popular waltz, *The Blue Danube*, as an example of the latter.) However, there is little or no difference between the cost of a CD of 'sophisticated' music and the cost of a CD of 'low-brow' music.

The Veblen consumer is definitely an *active* consumer. First, (s)he is active in choosing items for consumption that will achieve the desired purpose: of demonstrating the distinction of the consumer. It is hard to see how a passive consumer could hope to achieve distinction. Second, (s)he is active in the process of consumption – because such consumption will only demonstrate distinction if it is distinctive.

How does a Veblen consumer react to innovation? Some product innovations – notably the more radical innovations – would appear to offer an opportunity to demonstrate distinction. However, the demand for distinction can be subtle and the reaction to innovations more complex than appears at first sight. Thus, for example, in an analysis of the demand for distinction through the ownership of prestige cars, Swann (2001a) identifies two types of Veblen consumer: one who seeks distinction in novelty and one who seeks distinction in antiquity. The former is very likely to covet the latest model of Ferrari, for example. But the latter is more interested in a Rolls Royce Silver Cloud from 1955 than the latest model of Rolls Royce. For the former, innovation is welcome as a new source of distinction. For the latter, innovation is a way of ensuring that the brash *nouveau riche* city trader who spends his/her bonus on a new Rolls Royce cannot hope to match the timeless elegance of the Veblen consumer's 1955 Silver Cloud.

MARSHALL CONSUMER

Our fifth consumer is arguably the most active consumer of all. We shall call him/her, Marshall's consumer, after the great pioneering economist, Alfred Marshall.

In fact, Marshall was one of the originators of formal neoclassical theory of consumption – i.e. the economic consumer described above. But the early discussion in his *Principles of Economics* describes a much more interesting type of consumer. An important characteristic of Marshall's consumer is that the *way* in which (s)he achieves higher 'utility' may change significantly as the target rises. Marshall (1920, p. 86) recognised this in an important passage about the consumer:

> every step in his progress upwards increases the variety of his needs together with the variety in his methods of satisfying them. He desires not merely larger *quantities* of the things he has been accustomed to consume, but better qualities of those things; he desires a greater choice of things, and things that will satisfy new wants growing up in him.

Marshall's consumer becomes more subtle and varied in his/her consumption (Marshall, 1920, p. 86):

> As ... Man rises in civilisation, as his mind becomes developed ... his wants become rapidly more subtle and more various; and in the minor details of life he begins to desire change for the sake of change.

This is not simply an emergent demand for variety: Marshall's consumer becomes more social and conspicuous. Marshall is struck by the quote from Senior, listed above. Despite this, Marshall's consumer is selective in those areas in which (s)he seeks distinction. What starts as a demand to enable Marshall's consumer to take part in some 'higher activities' may in due course turn into a demand for more conspicuous purposes. Moreover, Marshall's consumer will not be satisfied with distinction alone. In due course (s)he aspires to excellence for its own sake, even in private consumption (Marshall, 1920, p. 89): 'For indeed, the desire for excellence for its own sake is almost as wide in its ranges as the lower desire for distinction.'

Marshall goes on to argue that when consumption progresses beyond its simplest forms, the wants of Marshall's consumer are driven by his/her activities, and not vice versa. This is a very important observation because it demonstrates the cumulative development of tastes through earlier consumption experiences. Moreover, let us remind ourselves of the following quote from McCulloch, repeated by Marshall (1920, p. 90), which we encountered in Chapter 2:

> The gratification of a want or a desire is merely a step to some new pursuit. In every stage of his progress he is destined to contrive and invent, to engage in new undertakings; and when these are accomplished to enter with fresh energy upon others.

This last quote, in my view, gives the best definition of a Marshall consumer. (S)he is an innovator, *actively* seeking out new and different consumption opportunities, and *actively* making the most of each consumption experience.

How does a Marshall consumer react to innovation? In general a Marshall consumer is as receptive to innovations as any consumer can be. The Marshall consumer is one who, even if (s)he had no apparent need for an innovation when it is first marketed, would want to try it out because (s)he is always interested in the 'new pursuit' and to 'engage in new undertakings'. This can mean that the Marshall consumer plays a pivotal role in developing the market for some new (and especially radical) innovation.

However, the Marshall consumer is not just important as an enthusiastic recipient of innovation. (S)he is, in a sense, an innovator him/herself: an *innovative consumer*. This is in line with the recent work of von Hippel (2005) on democratic innovation, which recognises that some innovations are very much consumer led rather than producer driven. It is useful to conclude this section with a brief review of some innovative aspects to consumption.

Studies of the use of new high technology products have shown how consumers do not necessarily make use of products in the way that was intended by the producers thereof. Consumers may frequently make unintended or unforeseen applications – and these must classify as innovative behaviour on the part of the consumer, because the consumer is not simply following an instruction manual for a new product. One example of this is the preference for texting over voice calls amongst teenage users of the mobile phone, which was not what the phone manufacturers and operators originally expected.

Household production theory treats the household utility-producing machine as a production process: the household combines goods and services with family labour to produce enjoyment. Thus, for example, take the case of the household chef's new recipe which reduces the cost of preparing a meal, or improves the quality of the meal produced from given ingredients. This would count as a consumer innovation.

When a consumer collects stamps, (s)he takes items that are often of modest value, combines them in an original and interesting way, and creates a stamp collection of some value. The same logic applies to other forms of collecting, where the aim is to create a whole which is worth more than the sum of its parts. This would include collections of books, coins, antiques, memorabilia, and so on. Indeed, in some cases, such as train-spotting, the 'items' that are collected are of almost no value in themselves, while the collection is all.

Another group of examples are where the consumer turns objects of no value (often waste objects) into something of value in use – even if not of market value. Thus children can often create new games by combining old toys and other objects in a new and innovative way. Some resourceful householders have made original furniture from waste items of wood, metal, and even glass bottles. Some rather stark examples of this are found in the study of sub-cultures. The aim of style innovation is to use everyday objects in a new and original way, and to imbue these objects with a new meaning. Members of the sub-culture use style innovation to reinforce their membership of the group.

In all of these cases, the consumer is being innovative by creating enjoyment or added value by combing objects in a new way. The objects may or may not have some value on their own, but the value of the whole is greater than the sum of its parts. The parent manages to keep the family happy at low cost. The collector creates a collection of value from items that are unremarkable on their own. The fashionable youth achieves a style to impress his/her peer group at low cost. Looking at these from a cost function perspective, all these forms of innovation offer cheaper ways of reaching a given level of welfare from a given price list, and as such are analogous to the process innovation.

ROUTINE CONSUMER

Our sixth consumer has been left to last because it is easier to define the character of this consumer as the negation of the rest. The routine consumer, described by Warde (2002), is perhaps the most passive of all. (S)he does not actively compute optimum consumption bundles like the economic consumer: instead (s)he just sticks to familiar routines in consumption. (S)he is not unduly influenced by advertising, in contrast to the Galbraith consumer. (S)he will probably not change his/her consumption patters even if members of his/her peer group change their consumption. (S)he does not seek distinction, in contrast to the Veblen consumer. And (s)he is definitely not an innovator, in contrast to the Marshall consumer.

The routine consumer just buys the familiar and ordinary. (S)he is hardly an active consumer. Indeed, (s)he is arguably an even less interesting character than the economic consumer. Moreover, of all the consumers we have described (s)he is the least interested in innovation: indeed, (s)he will be highly suspicious of innovation.

In describing the routine consumer, Warde (2002) said that his main message was that there must be more to the sociology of consumption than Thorstein Veblen. While conspicuous consumption does exist, it is not the norm.

NOTES

[1]   Parts of this chapter draw heavily on Swann (2002b).
[2]   Duesenberry (1960, p. 233), here quoted from Becker (1996, p. 17).
[3]   The marketing literature instead makes a distinction between *informative* and *transformative* advertising. The latter is directed at transforming the consumer's impression of a product or service.
[4]   Douglas (1983, p. 45), here quoted from Becker (1996, p. 13).
[5]   Smith (1776/1904a, p. 173).
[6]   Here quoted from Marshall (1920, p. 87).
[7]   Here quoted from Jevons (1878, p. 103).

16. The diffusion of innovations

Alongside the literature on models of demand and consumption there is a related literature on the diffusion of innovations across a market or an economy. Whereas the literature on demand and consumption seeks to understand the factors that influence the consumer's decisions, the literature on diffusion seeks to understand the rate at which consumers take up an innovation. The two issues are related – conceptually at least. For the rate at which an innovation diffuses will depend on the factors that influence individual consumption decisions and the rate at which these change over time. Having said that, the two literatures have grown independently because the linkages between them are rather complex. In this short chapter, we shall give a very brief overview of some of the main models of diffusion.[1]

In Chapter 12, we quoted Thomas J. Watson's famous under-estimate of the market potential of the computer. Projecting how the market for a new technology will grow is probably one of the most difficult questions in economics. Why is it so hard? There are several reasons. First, it is customary to assume that tomorrow's markets are just more developed versions of today's, but in fact new needs, applications, wants and sources of competitive advantage may be evident tomorrow that are not present today. Second, we often fail to recognise how 'path dependent' is the evolution of demand for a new technology – meaning that the minor historical details and events of today can have a decisive role in influencing the potential future market for the technology. And third, demand projections will depend on the evolution of this technology, but also of its rivals, and the latter is not always easy to foresee. Moreover, one of the regular features of competition in new technologies is that new rivals emerge from unexpected directions – especially when technologies are *converging*. This makes it very hard to know what will be the most important technological trends that influence demand for a particular technology.

The models that follow cannot cater for all these complications. But they represent some of the most influential models in the literature at present.[2]

'EPIDEMIC' MODELS

These models recognise that there is a certain similarity between the diffusion of a new technology amongst a population of users and the spread of an infectious disease amongst a population of people who do not have resistance. The epidemic model assumes that each time the consumer is 'exposed' to the new product or service there is a certain probability that (s)he will be 'infected'. In its simplest possible format, the epidemic model assumes (as with the infectious disease) that the rate of new cases (of adoption) is proportional to the *product* of the number infected (who use the technology) and the number who are not infected, but could be (the potential future adopters). Mathematically speaking, that is:

$$\text{New consumers} \quad \alpha \quad \text{Infected * Uninfected} \qquad (16.1)$$

We can solve this differential equation to give the familiar S-shaped diffusion curve, showing how the number of adopters increases over time. Simple mathematical considerations show that the rate of diffusion is fastest towards the middle of the diffusion process, and in the simplest version of the model the fastest rate of adoption per period is when exactly 50 per cent of the target population are infected.

In this form, the epidemic model makes some rather strong assumptions about 'infection' – i.e. how the adoption of the innovation will spread. First, each exposure increases the buyer's probability of purchase regardless of who conveys the message. But in practice, buyers tend to be more heavily influenced by those that they know and trust – for example, those in the same peer group (or, for industrial buyers, those in the same industry). As a consequence, epidemic diffusion may be faster *within* social/industry groups than *across* groups. Some models recognise that more complex patterns of interaction are required before a user will 'catch' the technology from others. These more subtle epidemic diffusion models generate slightly different results: diffusion curves are still S-shaped, but not necessarily symmetric.

Second, each exposure increases the buyer's probability of purchase regardless of the quality of the pioneer's experience. But in practice, this is too simplistic. The social process which spreads the use of a technology is taken to be the exchange of information amongst potential users, and the pioneer plays an important role as *demonstrator* and *educator*. In more subtle models of this sort, it is not just the *number* of pioneers that count, but also how influential each one is. Moreover, the effect that the pioneer has on subsequent adopters is also dependent on his/her experience with the technology: good experiences promote diffusion, while bad experiences retard diffusion.

In the very long-term, the level of diffusion approaches (asymptotically) a maximum or *saturation* level of diffusion. This technique is useful for projecting how fast the uptake will be, given a certain saturation level of diffusion, but it does not *in itself* help us to estimate the saturation level.

The epidemic model remains relevant in a variety of settings. It is relevant to the understanding of diffusion when price and quality are not an issue, and when consumers are not constrained by income or other factors. Speeds of diffusion will vary enormously from one technology to another – compare the diffusion of the car and a particular pop record, for example.

PRICE AND QUALITY EFFECTS

In this approach to modelling diffusion, it is assumed that the main driving force behind the diffusion of a new technology is that as it becomes cheaper and/or better, ever more buyers are willing to adopt. Or, in short, as price falls and quality improves, more consumers enter the market.

This approach is most relevant for products subject to rapid technological change, so that their price falls or their quality improves. Moreover, this approach is based on the underlying assumption that we have an *economic consumer* – as discussed in Chapter 15. As a result, we can analyse the way the consumer responds to price reductions and quality improvements using the *product map* and *territory map* introduced in Chapter 5. To recap, the territory map describes the sorts of consumers that buy a particular product. Any price reductions and quality improvements in a particular product extend the territory of that product.

The easiest way to envisage this process at work is in a pair of diagrams. Figure 16.1 summarises the trend in one of the simplest measures of improvement in a technology: the cost per component. The (hypothetical) line drawn here is related to Moore's Law: as originally stated, this observed that the rate of increase in the number of components per semiconductor chip was doubling every year, though the rate of advance started to tail off as the technology matured. This also implies an exponential decline in the price per component. Then Figure 16.2 draws out the implications for buyer demand. It shows the numbers of buyers that are prepared to pay different prices per component. As the price per component falls over time, so the cut-off point sweeps leftward thus drawing ever more users into the market.

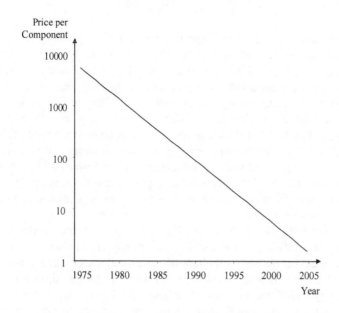

Figure 16.1 Rapidly declining prices per component

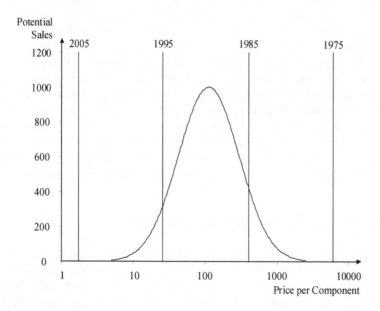

Figure 16.2 Distribution of willingness-to-pay

The number of users at any time is the area under the density curve to the right of the vertical line in Figure 16.2, which simply marks where the technology has reached in a particular year. In 1975, almost no buyers are willing to pay the going rate for the technology. In 1985, only buyers on or to the right of the vertical line marked '1985' are willing to pay the going rate for the technology, so demand is limited. By 1995, however, the cost of the technology has declined markedly, and all buyers on or to the right of the line marked '1995' are willing to pay the going rate, so demand is high. And by 2005, the price per component has declined to a level where all buyers for all applications can afford the technology, so price is not a constraint on demand any more. Careful inspection shows that this process generates an S-shaped diffusion curve as before.

The speed of diffusion here depends on the speed of price decline and the variance of the willingness-to-pay of different buyers. As drawn, the variance is very large, but so also is the rate of price decline. As noted above, these rates of change provide a more rapid impetus to the diffusion of the technology than rising incomes or turnover. As Figure 16.2 is drawn, in 20 years the technology moves from almost zero diffusion to very widespread diffusion.

To apply this technique for making projections requires some knowledge or estimates of the density of users willing to pay different prices. By definition one cannot know in a particular year the density of buyers to the left of the line marking the 'state of the art' in that year, as they have never revealed themselves. On the other hand a number of market research techniques exist to estimate the likely density of consumers in this area.

Sometimes this simple picture of declining price per component is an over-simplification. It ignores the fact that it is not just the decline in the price per component that draws more buyers into the market, but also the availability of products and technologies of higher quality than was previously available. This theme has been discussed in Chapter 5. The rate of diffusion in this process, again, depends on the speed at which prices fall and quality improves, but also on the density of buyers in the newly opened territories.

INCOME, SIZE AND OTHER DRIVERS

Another force leading to diffusion is related to growing income, turnover or other characteristics of the buyer. As income per capita rises gradually over time then consumption of some products will rise. At an early stage, buyers may be limited to those on relatively high incomes, but as the average level of incomes rises, more and more buyers enter the market. The same principles can be applied to the diffusion of new technologies amongst firms. For some expensive new technologies, early buyers may be the larger firms (in a particular industrial sector), but as the smaller firms grow, they too would be potential buyers.

And this basic approach can also be used with other consumer or firm characteristics: as consumers become educated they are more likely to buy certain products; as firms become more concerned to seek new sources of competitive advantage in an ever more competitive market, they may be more likely to buy into some new technologies. In the literature, this approach is described as a *probit* approach to diffusion – named after the econometric technique used to model the probability of adoption as a function of characteristics of the customer.

As before, this process generates an S-shaped diffusion curve. But, in general, diffusion promoted by growth in incomes, revenues or other buyer characteristics tends to be a slower process than diffusion attributable to other effects. The reason for this is simply that the rate of growth of incomes is typically quite a small percentage rate. GDP per head rises at perhaps 3 per cent per annum on average, and most consumers' incomes rise no faster than that. Large companies do well to grow any faster than that. Some smaller companies of course can grow a good deal faster, but they are outnumbered by companies that don't.

The key difference between this approach and that in the last section is that here we model adoption as a function of customer characteristics while, in the last section, we modelled adoption as a function of changing product characteristics and price. The approach of this section is particularly relevant to study the diffusion of products for which price and quality are stable, and where consumers are well informed about product from the start.

STRATEGIC INCENTIVES FOR ADOPTION

The models of diffusion described above contain no element of strategic interaction. Game theory models of diffusion, however, suggest that such strategic interaction between players may be an important factor in explaining the rate of diffusion. For example, some adoption decisions may be driven by a desire to forge ahead of competitors in order to achieve a cost, quality or performance advantage. Equally, some adoption decisions may be driven by a need to catch up with competitors and thus to make good a cost, quality or performance disadvantage. Interestingly, the former incentive may be strongest when diffusion has not progressed far in the market: in that case, there is still scope to forge ahead. But the latter incentive may be strongest when diffusion is quite well progressed: in that case, the need to catch up is most pressing. The rates of diffusion generated in these models can be very rapid.

Karshenas and Stoneman (1993) made an important methodological breakthrough in the empirical analysis of strategic models of diffusion, although in the particular example they studied, such strategic interaction was not especially important. These models of strategic interaction may be relevant when information, price/quality and income effects are not important factors in explaining the rate of diffusion.

BANDWAGON EFFECTS

In the previous three sections, a company's or consumer's decision to adopt is assumed to be independent of what others decide. The recognition that both for individual consumers and for industrial buyers this need not be so opens up an important strand of diffusion modelling.

We start with the case where customer A's adoption of an innovation makes customer B *more* likely to adopt. This could be the result of strategic interaction as just described or a result of *network effects* (see Chapter 7 for a recap on this topic). It may be that there are direct network benefits from adopting a technology which has a large installed base of users. Or alternatively, the existence of a large installed base is often a powerful signal to some buyers that this technology is no longer the risky prospect that it appeared at first.

This is sometimes called a *penguin effect*. The reason for this metaphor is that penguins often line up along the edge of the ice waiting to jump in to catch fish. But they know there is a risk of being the first to jump in because there is a danger from predators. If one penguin, wisely or otherwise, has the nerve to jump in first and is not attacked by predators, then the rest will soon follow as the perceived risk has fallen. This is another example of the 'fast second' concept, described in Chapter 2.

This feature makes for a very powerful and rapid diffusion process which is sometimes described as a *bandwagon*. If one technology gets ahead of the field, the market will quite rapidly lock into that technology. This approach is used to model the emergence of *de facto standards* or *dominant designs* in markets (see also Chapter 7).

This feature of rapid take-off can also arise where firm B faces competitive pressure to adopt a new technology because their rival firm A is using it with good effect to take market share away from B. While firm B may not be especially interested in adopting the technology on its own merits, they may become so if firm A exploits the technology for competitive advantage. This can lead to a 'me too' attitude to adoption, and a very rapid rate of diffusion.

This process of diffusion as a result of competitive pressure, operating in addition to those described in the first three sections above can further accelerate the process of diffusion. One major feature of technological change noted above is that firms tend to find new competitors emerging from new and unexpected directions. That means the technology adoption decisions of an ever greater number of firms can impinge on the decision faced by one particular firm. This proliferation of competitors can make diffusion resulting from competitive pressure a very rapid process, and will continue to raise the level at which the market reaches saturation.

DISTINCTION EFFECTS

Now we turn to the case where customer A's adoption of an innovation makes customer B *less* likely to adopt. Perhaps the customer has a desire for distinction or the value of adoption may actually fall if too many others have already adopted. We saw in Chapter 15 that demand may be as much influenced by the behaviour of the *distinction group* from whom the consumer wishes to distinguish him/herself, as by the behaviour of the *peer group* with which (s)he wishes to associate. When this feature is important, it tends to suggest that there are likely to be clear upper limits to diffusion, covering far less than 100 per cent of the potential population. Moreover, when this feature is important, it can generate waves or cycles in the popularity of a product (Cowan et al., 2004).

These effects can also arise in the context of industrial buying if a group of pioneering buyers is (for whatever reason) seen as a 'distinction group' by another large set of industrial buyers. This could happen if early applications in some sectors were generally thought to be unsuccessful, and if other companies seek to distinguish themselves from the *geekish* early users. This emphasises again how important it can be that early applications of a technology are successful to maximise its ultimate diffusion.

CONCLUSION

In most practical cases, actual diffusion patterns will exhibit a mix of the different phenomena described above. Moreover, it is often the case that several of the different consumer types described in Chapter 15 will play a distinctive role in the observed pattern of diffusion. Marshall consumers (and perhaps also Veblen/Bourdieu consumers) may be essential as pioneers at an early stage, while Douglas consumers become more important later on, and Galbraith consumers are most important as the market approaches maturity.

NOTES

1. Parts of this chapter draw heavily on Swann (2002b).
2. Pioneering works on diffusion are those of Griliches (1957), Mansfield (1961) and Rogers (1995). The Appendix lists some surveys of the literature.

PART V

The effects of innovation

17. Innovation and trade

The aim of this chapter is to give a very brief overview of some of the ways in which innovation can impact on trade patterns. While it might be convenient if we could ask, 'what does trade theory say about innovation?', the question has no simple answer because there are many generations of trade theory and they do not all give the same answer. Indeed, each of these gives a rather different perspective on innovation.

We could argue that trade theory started with the classical theory of absolute advantage – as in Adam Smith's work, for example. According to this, a country would be an exporter in any industry in which it had an absolute advantage. This meant that any country with an absolute disadvantage in every industry would be unable to export at all. The subsequent theory of *comparative* advantage, first articulated by Ricardo and developed by Hecksher and Ohlin (Ohlin, 1933), took a very different view. It asserted that countries would be exporters in those industries in which they had a comparative advantage – even if they had no absolute advantage in any industries. This has altogether less pessimistic implications.

While this Hecksher-Ohlin theory has been very influential in the history of economic thought, its predictions were shown to be wrong. Several theories were advanced to explain these contradictions – notably the product life cycle theory of Hirsch (1965) and Vernon (1966) and the technology gap theory of Posner (1961). These are of particular interest to the student of innovation.

Subsequent developments in trade theory include the theory of intra-industry trade (Grubel and Lloyd, 1975) which attempts to explain why a particular country may be an importer *and* an exporter of a particular category of products. Strategic trade theory describes how a country may use tariffs and subsidies (including export subsidies) to alter the balance of international competition in its favour. And, of particular interest in the context of this book, the work of Krugman (1991) interprets trade patterns as the result of industrial location decisions.

In this chapter, we shall focus on four areas of theory which give a particular role to innovation. We have already, in Chapters 3, 4 and 5, made a distinction between process innovation and product innovation. We need to bear that in mind in what follows because in some theories at least, the implications of product and process innovation are different. Moreover, one other type of innovation plays an especially important role in determining trade patterns. Innovations in transport and communications which reduce the cost of trading over distance have perhaps the most significant impact of all.

THEORY OF COMPARATIVE ADVANTAGE

In this section, let us consider the potential trade between countries A and B in products X and Y. According to the theory of absolute advantage, country A will be an exporter in those industries in which they have an absolute *advantage* over country B, and A will be an importer in those industries where it has an absolute *disadvantage* compared to B. This means that if A has an absolute advantage in X but an absolute disadvantage in Y, then A will export A and import Y. And it also means that if A has an absolute advantage in X *and* Y then it will export *both* X and Y and country B will be unable to export *either* X or Y. This theory of absolute advantage has a rather pessimistic implication for backward countries.

According to the theory of comparative advantage, by contrast, what matters is not whether countries have an *absolute* advantage or disadvantage. What matters is *comparative* advantage. Suppose that country A has a large advantage over B in producing X, but A has only a small advantage over B in producing Y. In that case we can say that A has a *comparative advantage* in X while B has a *comparative advantage* in Y – even though B is at an *absolute* disadvantage in Y. According to the theory of comparative advantage, A will be an exporter of the product in which it has a comparative advantage (X) while B will be an exporter of the product in which *it* has a comparative advantage (Y).

This makes sense because this allocation of production maximises the total level of production and therefore combined national income. A maximises its income by specialising in production X and buying that level of Y it wants by trading with B. The theory of comparative advantage gives exchange rates the role of ensuring that markets for the two traded goods will clear.

The Heckscher-Ohlin model took this further by showing the relationship between comparative advantage and each country's endowments of factors of production (capital and labour). In particular, the Heckscher-Ohlin theorem stated that nations would specialise in producing commodities where the production process uses their relatively abundant factor intensively. The implication of this would be that the USA (as a capital rich country) should export commodities whose production is relatively capital-intensive, and import commodities whose production is relatively labour-intensive.

Where does innovation fit into this theory? Our answer is in two parts.

First, a process innovation adopted in one country (A) which reduces the cost of producing X but which is not yet adopted in B, may shift A's balance of comparative advantage towards product X and away from product Y. In short, process innovations can change the balance of comparative advantage, and hence change trade patterns, especially when diffusion across different countries proceeds at different rates.

Second, an innovation which reduces the cost of transport and/or communications can have quite subtle effects on trade patterns. First of all, if we start from a position where transport between A and B is so costly that trade is just not viable, then a transport innovation which reduces transport costs may be enough to start up trade between the two where there was none before the innovation. Moreover, if one good is heavy and costly to transport while another is light and cheap to transport, such a transport innovation may also change the balance of comparative advantage and hence change trade patterns. (To show this requires a somewhat complicated model which goes beyond the scope of the present text.)

Product innovations, however, don't really fit into this theory because the theory is based on the assumption of fixed products (in common with almost all economic theory of that time).

PRODUCT LIFE CYCLE THEORY OF TRADE

The Heckscher-Ohlin theorem has been exceptionally influential in trade theory. However, Wassily Leontief (1953, 1956) found that US trade patterns were not actually consistent with the predictions of the Hecksher-Ohlin theorem. Leontief compared the capital and labour requirements of US exporting industries and also those US industries producing commodities in which the US is a net importer. He found that some US exporting industries were more labour-intensive than US import-replacing industries. This famous finding has ever since been known as the *Leontief paradox*.

The product life cycle model of trade (Hirsch, 1965, and Vernon, 1966) and the related technology gap theory (Posner, 1961) were advanced as theories that could explain the Leontief paradox.

The essence of the product life cycle model is that during its evolution an industry passes through four main stages of development – introduction, growth, maturity and decline – and that the nature of competition changes in important ways as the industry passes through these stages.

During the introduction stage of the life cycle, growth is relatively slow. This may be: (1) because of buyer inertia or ignorance; or (2) because there is not a large enough group of pioneering consumers to demonstrate the potential of the new product and start the *bandwagon*; or (3) because prices are high and the product is still at an experimental (and perhaps unreliable) stage. Growth may also be slow if buyers wait to see what will become the *industry standard*.

In the growth stage, sales growth is very rapid because buyer inertia has been overcome and the bandwagon is rolling, and also because prices are falling and the product has become more reliable. In cases where *industry standards* are important, the rapid growth in consumer demand may also reflect the fact that consumers are clearer about what this *industry standard* will look like, and so find the product a less risky buy.

In the maturity stage, diffusion throughout the target group of consumers is reaching saturation, so that growth starts to level out and then fade away. The level of industry sales depends on rate of growth of the relevant buyer group, and the frequency of replacement (or repeat) purchase amongst consumers who bought at an earlier stage.

Finally, during decline, sales fall off partly because of the pattern that emerged in the maturity stage, but more so because of the fact that new products and technologies are starting to appear that are offer a better price/performance trade-off than the present product.

Perhaps the three most important trends here are these. (1) Product manufacture often moves from being small scale, skill intensive and experimental at the introduction stage, to large scale, low skill, and standardised at the maturity and decline stages. (2) Early consumers are pioneers, often relatively unconcerned about prices, and willing to put up with a degree of unreliability in the product, while consumers at the end of the product life cycle are either sophisticated buyers who demand reliability at a reasonable price, or are very price conscious. (3) Competition from other firms is low during introduction, but grows considerably during the growth stage. The number of competitors starts to fall off during maturity, and even more so during decline, but the intensity of price competition is undiminished.[1]

On these foundations, Hirsch (1965) and Vernon (1966) built a very compelling account of trade flows. Their argument is that the USA would tend to specialise in exporting products which are in the early or growth phases of their life cycles but would tend to import products which are in the maturity or decline phases. By exporting the former categories, US firms can sell in markets where technological advantage is paramount, and there they surely have an advantage. And by importing the latter categories, they avoid trying to compete in markets where cost considerations are paramount – and where they cannot hope to compete against companies based in the NICs.

This theory can (to some degree at least) explain the Leontief paradox. For products in the early stage of the life cycle tend to be produced by relatively skilled-labour-intensive processes, while products in the maturity and decline stages of the life cycle tend to be produced by relatively capital-intensive processes. This explains the apparently anomalous observation by Leontief: US producers in export-intensive industries will have relatively skilled-labour-intensive production processes while US producers in import-intensive industries will have relatively capital-intensive production processes.

Where does innovation fit into this theory of trade? We would say that innovation lies at the heart of this theory because the product life cycle theory is, in a way, a theory of innovation – in particular, a theory of how the balance of innovation alters over the life cycle. In particular, a typical proposition of life cycle theory is that product innovations are most important in the early and growth stages while process innovations are most important later in the life cycle. Indeed, we could say that the incidence and frequency of product innovations versus process innovations is one measure of where the product is in its life cycle. If product innovations predominate, then the product is at a comparatively early stage of the life cycle and a high-technology country will tend to be an *exporter* at that stage. But if process innovations predominate, then the product is at a comparatively late stage of the life cycle and a high-technology country will tend to be an *importer* at that stage.

Moreover, innovations in transportation and communication can impact on trade flows. If transport costs are prohibitive, there will be no trade between A and B, regardless of life cycle considerations. But if transport innovations reduce transport costs somewhat, then trade will start to take place in products at either end of their life cycles (very early or very late). And if transport costs fall more or less to zero, then an even larger class of products will start to be traded.

INTRA-INDUSTRY TRADE THEORY

Another anomaly of the theory of comparative advantage is that it predicts all production in a particular industry will be concentrated in one trading partner. In terms of our earlier example, country A specialised in X and country Y specialised in Y. But in practice, it is very common to find that country A may be both an exporter and an importer of the same product (X). So, for example, the UK is both an importer and an exporter of refrigerators and washing machines.

The intra-industry theory of trade was advanced to explain that phenomenon. The theory stresses the role of product variety and economies of scale. Because of economies of scale, countries specialise in particular parts of the product spectrum. So, returning to our example of refrigerators, we find that: (a) low-cost products tend to be imported from the countries of former Eastern Europe; (b) mid-priced products are often produced in the UK and also exported; and (c) the high-price products are imported from Germany and Scandinavia.

Intra-industry trade becomes more common the wider are our product categories. So, for example, if we look at trade in 'white goods' as a whole (i.e. refrigerators, washing machines, dish-washers, etc.) we find a lot of intra-industry trade; if we look at trade in refrigerators alone we still find it, but not so much; and if we look at a very narrow segment of the market for refrigerators (e.g. the most expensive models) there is much less intra-industry trade. To put it another way, if there is a lot of product variety within one statistical category, we may see a lot of intra-industry trade.

Where does innovation fit into this theory? Process innovations that increase economies of scale will tend to increase intra-industry trade because such innovations make it more important for each trading nation to specialise in particular segments of the market. And, again, as in Chapter 16, process innovations that diffuse at different rates in different countries will change the balance of trade.

Product innovations also fit neatly into this theory. Countries may specialise in particular product innovations in particular market segments where they have a trading advantage. Moreover, product proliferation (a special form of product innovation, discussed in Chapter 5) will tend to lead to more intra-industry trade as proliferation creates more market niches and therefore more opportunities for specialisation.

Innovations in transport which reduce the cost of trading will also increase intra-industry trade. If transport costs are high, then traded goods are too expensive even if one country enjoys an advantage in a particular market segment. In that case, low-priced and high-priced products are all domestically produced. But as transport innovations reduce the cost of trading, they make it ever more likely that a specialist producer in one country can undercut producers in another country. And when transport costs fall towards zero, it is quite likely that each trading nation will specialise in supplying just a few market segments and import all (or most) products in other segments.

TRADE AND LOCATION

Finally, we should mention Krugman's (1991) argument that trade patterns are the result of industrial location decisions. This theory has something in common with Porter's (1990) revival of interest in industrial clusters (see Chapter 13).

Indeed, our discussion of clusters in Chapter 13 gives an informal account of this perspective – and, in particular, the effect of transport innovations on trade. In that chapter, we saw that if transport costs are high then there is little trade between regions. But as transport costs decline, production becomes more concentrated in one cluster, and inter-regional trade increases. If we replace 'the region' by 'the country' then we have in effect a theory of trade.

This theory seems highly relevant to trade in electronics, as discussed in Chapter 14. There we saw that as transport and communication costs continued to fall, different industry clusters would specialise in particular electronic components, and global production of these would become very concentrated in just a few clusters. A computer would consist of components made in many different countries and assembled somewhere close to the final market. Along with this increased specialisation and concentration of production in clusters, we would find increased international trade in components and increased *component miles* within each computer. This is a point we shall return to in Chapter 22, when we discuss the impact of innovation on sustainability. And any process innovation that increases economies of scale will reinforce this tendency towards global concentration of production in a few clusters.

We conclude this chapter by taking a look at the *death of distance* hypothesis from Cairncross (1997). She argued that this *death of distance*, meaning declining transport and communication costs, could lead to:

1) 'The fate of location'
2) 'The irrelevance of size'
3) 'Less need for immigration and emigration'
4) 'Near frictionless markets'
5) 'More global reach; more local provision'
6) 'The loose-knit corporation'.

It could be argued that some of these predictions are right but some are wrong. In particular, I would argue that (1), (2) and (3) are probably wrong while (4), (5) and (6) are probably right. Why do I argue that there is no 'fate of location'? Because in global market, survival of any global player depends on it exploiting all the advantages offered by the cluster(s) in which it is located. Location becomes even more important for that reason.

Why do I argue that there is no 'irrelevance of size'? Because space-shrinking technologies increase the extent of the market. If economies of scale continue to be important then this reduction in transport and communication costs makes it ever more likely that the global market in any industry will be dominated by a few very large players. Small players will have a role, but it will be a very specialised role in specialised niches.

And why do I dispute the assertion that there will be 'less need for immigration and emigration'? Because if, as discussed in Chapter 13, falling costs of transport and communication lead to more concentration of business in clusters, then some of those located in the hinterland will find that they face declining employment prospects at home and will increasingly be drawn to emigrate to these large clusters.

Paradoxical as it may seem, declining costs of transport and communication may make location more (rather than less) important, may make size more (rather than less) important and may make emigration and immigration more (rather than less) important.

NOTES

[1]   Klepper (1996) provides a more recent synthesis of research on the product life cycle.

18. Innovation and market structure

Most economists would agree that the relationship between market structure and innovation runs in both directions: market structure influences innovation but equally, innovation influences market structure.

There is a large and quite old literature on the question of how different market structures will generate different patterns of innovation. Equally, it is well recognised that the innovative activities of different firms will impact on market structure. This latter linkage can happen either because innovative activity influences performance (and hence structure) or because of the connection between innovation and entry: often innovations by incumbents may act as a barrier to entry, but at the same time, especially in the formative stages of a market, innovation may come from new entrants.

While the two causal chains – from structure to innovation and from innovation to structure – can be separated conceptually, and sometimes empirically, any discussion of one almost inevitably has to make some reference to the other. In theoretical studies of strategic innovative behaviour, the two are interconnected because, for example, it is the knowledge of the feedback from innovation to concentration (via entry barriers) that provides the strategic incentive for incumbent monopolists to innovate. More generally, one causal link becomes part of the explanation of the reverse link.

Moreover, the two causal linkages are interconnected because of the closed system dynamics that they generate. If concentration has a positive effect on innovation and if innovation reinforces concentration, then in response to an exogenous increase in innovative activity we find a virtuous circle of continuing positive feedback, with increased concentration and increased innovation. Conversely, if concentration has a negative effect on innovation, and innovation has a negative effect on concentration, then in response to an exogenous increase in innovation we find a spiral of reduced concentration and increased innovation.

In this chapter we shall give a very brief account of the relationship from market structure to innovation and then a rather more detailed account of the reverse relationship, because we feel that is the more interesting and complex of the two. In the final section we focus on a more specific question about the effect of a particular innovation on market structure: will the growth of the Internet and e-business make markets operate as if there were perfect competition? While some believe that will be the effect of the Internet, there are some good reasons to be cautious about such an assessment.

FROM MARKET STRUCTURE TO INNOVATION[1]

Our objective here is to summarise the main conclusions that have emerged from the large literature on this topic. The list of suggested readings in the appendix includes some detailed surveys of this literature.

At its simplest, the consensus seems to be that the effect of market structure on innovation depends on the dual forces of incentive *and* opportunity. By *incentive*, we describe the extent to which a firm in a particular market structure has an incentive to innovate. By *opportunity*, we describe the extent to which a firm in a particular market structure has an opportunity to engage in innovation. This last will depend in large measure on the *profitability* of the firm because it is well recognised that it is hard to raise external capital to fund innovative activities, in part because innovation is so risky, but even more so because of a fundamental information asymmetry between the innovator and the lender. (We return to that issue in Chapter 22.)

The literature tends to assume that innovation requires *opportunity and incentive*. Now we can describe a contrast between perfect competition, monopoly and intermediate forms of oligopoly.

Under perfect competition, firms cannot make super-normal profits and this severely constrains their opportunity to innovate. The extent of incentive for perfectly competitive firms depends on whether perfect competition is a permanent condition or something that might be altered by innovation. If the former, then perfect competition provides no incentive to innovate either, because the innovator incurs costs without any hope of breaking free from that intense competition which erodes all his/her profits. If the latter, on the other hand, there is an incentive – at least to the extent that the innovator can change his/her company from a perfectly competitive one into a firm with some market power, at least. But either way, if innovation requires *opportunity and innovation*, then there can be little or no innovation in perfect competition because there is *little or no opportunity*.

Under monopoly, firms make super-normal profits and hence there is plenty of opportunity to innovate. But is there an incentive? The answer to this depends on whether the monopolist is a permanent monopolist – that is, there is no potential entry – or if (s)he is currently a monopolist in what might be a contestable market. If the former, and if (as we argue in Chapter 20) much of the firm's incentive to innovate is to enhance its competitiveness and its market share, then it is arguable that an unthreatened monopolist has little incentive to innovate as (s)he already has 100 per cent of the market. But if the latter, and if the monopolist cannot count on keeping 100 per cent of the market, then the monopolist would have an incentive to innovate just to see off potential competitors. Now, in this case, if innovation requires *opportunity and incentive*, then the amount of innovation under monopoly depends only on the extent to which the monopolist holds a permanent or contestable monopoly. If permanent, and there is no prospect of entry, then there is little incentive to innovate and therefore little innovation. If contestable, then there is a great incentive to innovate and also a great opportunity, so we would expect a lot of innovation.

In the case of oligopoly, we find that there is both *opportunity and incentive*. There is opportunity because the oligopolist makes some supernormal profits, even if not as much as the monopolist. And there is incentive, because (even in the absence of further potential entry) the oligopolist's market share cannot be taken for granted and needs to be defended against raids by existing rival oligopolists. As there is always opportunity and incentive, in this market structure, then there is always innovation. The outcome of these arguments can be summarised in Figure 18.1.

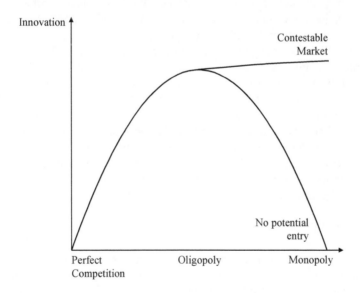

Figure 18.1 The relationship from market structure to innovation

The inverted-U relationship (the solid line) from market structure to innovation is a simple (but reasonably accurate) summary of what the literature has to say on the matter.

We have also drawn in an additional line showing the relationship between market structure and innovation in the case that any monopoly is *not* a permanent monopoly but a *contestable* one. This line reaches a slightly higher level of innovation under monopoly than under oligopoly. Why so? This happens because when a monopoly is not permanent but contestable, then the monopolist has a greater opportunity and a greater incentive to innovate than the oligopolist. The opportunity is greater because the monopolist has greater market power and can therefore secure greater profits. The incentive is greater because the threatened monopolist has more market share to lose (100 per cent indeed) than the threatened oligopolist (who by definition must have a market share below 100 per cent).

FROM INNOVATION TO MARKET STRUCTURE[2]

This section summarises some of the economic literature on the reverse linkage from innovation to market structure. We start with empirical evidence and then discuss theoretical evidence. We also survey some of the organisational literature on how firms cope with rapid innovation – because this is highly relevant to understanding this relationship.

The literature about the reciprocal relationship between innovation and market structure can perhaps be categorised along two axes. First, is it empirical or theoretical? Second, which discipline does it stem from: economics or organisational behaviour? Admittedly, this is an over-simplistic categorisation because a number of contributions are of both theoretical and empirical substance, and because there has been an increasing (and healthy) mutual exchange between the economics and organisation literatures. Nevertheless, it is a useful device for organising what follows. Three subsections follow: the first on empirical economics, the second on theoretical economics, and the third on organisational behaviour.

Empirical Evidence

In the empirical economics literature, two polar positions have emerged. Thus at one end it is argued that rapid technological change under difficult imitation conditions tends to increase concentration. Phillips (1956, 1966, 1971) was one of the first to expound and analyse this hypothesis. At the other end, Blair (1948, 1972) has argued that recent (post Second World War) innovations have tended to reduce minimum efficient scale and so have been deconcentrating.

Phillips (1956) and Horowitz (1962) were among the first to explore the reverse feedback from innovation to market structure. Phillips (1966) examined this with data on eleven industry groups and found that when the technology permits product change and differentiation, the scale of surviving firms is large and concentration is high. In his book on the aircraft industry, Phillips (1971) found that relative success in innovation was an important determinant of the growth of firms and of growing concentration.

Mansfield produced another early study of the relationship between innovative activity and subsequent firm performance, and again his conclusion was that innovative success could lead to faster firm growth and hence to greater concentration (Mansfield, 1962). Mansfield's (1968b) study of the steel and petroleum industries found that successful innovators grew much more rapidly than previously comparable firms for a period of five or ten years after the innovation. Mansfield (1983, 1984) examined some major process innovations in the chemical, petroleum and steel industries, and found that process innovations were more likely to increase (than reduce) economies of scale and hence would be a force leading to greater concentration. Among product innovations, on the other hand, the picture was less clear cut. In steel and petroleum, product innovations were more likely to increase concentration. In the chemical industry, on the other hand, just as many product innovations led to a reduction in concentration as led to an increase. And in the pharmaceutical industry, product innovations were more likely to reduce concentration.

This positive feedback is an example of a more general tendency that 'success breeds success', or what has been termed the Matthew effect (Merton, 1973). This manifests itself as a positive autocorrelation in firms' growth rates. Thus, for example, in a study of 2,000 firms in twenty-one UK sectors over the period 1948-60, Singh and Whittington (1975) found that firms with above average growth rates in the first six years also tend to show above average growth performance in the second six years.

Turning to the effect of innovation on market structure, it is recognised that innovation can generate barriers to entry and so lead to sustained concentration. Thus, Comanor (1964) found that expensive and risky R&D would act as a barrier to entry in the pharmaceutical industry. Stonebraker (1976) found that the risk of losses (through failure) to industry entrants was positively correlated with industry R&D intensity (R&D spend/total sales). Freeman et al. (1965) found that R&D was a barrier to entry in the electronic capital goods industry. Firms had to have strong development capacity to assimilate the inventions of others and to gain access to existing technology through cross-licensing, know-how agreements and patent pools.

R&D is a barrier to entry in the Mueller and Tilton (1969) analysis of industry dynamics over the product life cycle. At an early stage of the life cycle, neither absolute nor relative size may be essential for innovative success, but at a later stage (of technological competition) the large industrial research laboratory operates at an advantage, and if large R&D programmes are maintained, this will build a barrier to entry. Pavitt and Wald (1971) also conclude that small firms have their greatest opportunities in the early stages of the life cycle, when scale economies are lowest, but as the technology matures the opportunities for small firms decline.

Other literature casts light on this concentrating hypothesis in a different way. Menge (1962), for example, argues that rapid style changes in the car industry would impose greater costs on small producers than on large producers, and hence would be a force for concentration. This is one example of firms using rapid technology change as a device to raise rivals' costs (Salop and Scheffman, 1983). Following from Phillips' in-depth study of the aircraft industry, a number of other industry studies have lent weight to the argument that rapid technology change and high imitation costs are a force for increased concentration. These include studies by Dosi (1984) and Malerba (1985) on semiconductors, Katz and Phillips (1982) on computers, and Altschuler et al. (1985) on the automobile industry.

Now, we turn to those whose argument is that technological change has been deconcentrating. In an early paper, Blair (1948) argued that the previous ubiquitous trend towards increasing plant size and market concentration was being halted by certain fundamental shifts in technology. His later major study (Blair, 1972) comprehensively reviewed the impact of technology advance on scale economies. He argued that while from the late eighteenth century through to the 1930s technological change had been a force towards increased concentration, the position was changing as newer key technologies were having the opposite effect, reducing plant size and capital requirements for optimal efficiency. Over the earlier period advances in steam power, materials and methods of production and transport had acted to increase scale economies, while in the latter period developments in electricity, plastics, trucks, materials and production methods had served to reduce scale economies.

Relatively few econometric studies have provided evidence in support of this perspective. One is that of Geroski and Pomroy (1990), which estimated a dynamic model of market concentration from data on 73 UK industries over the period 1970-79 and found that innovations were deconcentrating. In another paper, Geroski (1990) found a negative effect of concentration on innovative intensity, and the two studies together imply a negative feedback model where innovation reduces concentration which further increases innovative activity. This is a very interesting contrast to the positive feedback models described above.

Economic Theory

Most of the economic theory that analyses how rapid technology change impacts on market structure tends to imply persistent dominance, or 'success breeds success', and so supports the thesis that rapid technology change is concentrating.

A well-known and early theory of how technology change reinforces social structures is that of Marx (quoted already in Chapter 1): 'The bourgeoisie cannot exist without constantly revolutionising the means of production'. Perhaps the next great theorist to address this was Schumpeter (1954), who saw to the heart of the matter. There are, by common consent, several Schumpeterian hypotheses. One hypothesis endorses the 'success breeds success' principle. Large firms and firms with market power can take advantage of such scale economies as there are in R&D (though these are uncertain, to say the least) and have the retained profit with which to finance R&D programmes. In addition, a companion Schumpeterian hypothesis notes that successful innovation reinforces market power, and the prospect of a (temporary) monopoly position consequent on innovation is one of the chief attractions of such innovative activity. Size and market power facilitate some aspects of (and components of) innovation, and innovation reinforces size and market power.

An important, and unjustly neglected, study by Downie (1958) provided a detailed economic analysis of why we should expect a systematic tendency for technical change to increase concentration. Firms which already have relatively low costs will be able to reduce their average cost faster than those that have relatively high costs, and so competitive price reductions by the low cost firms will drive out the relatively high cost firms. A similar mechanism has been used in some of the evolutionary models of technical change and market structure (see, for example, Nelson et al., 1976; Nelson and Winter, 1978).

The positive feedback in this model is observed in more recent theoretical studies. The simulation models used by Nelson and Winter (1982) exhibit a rich diversity of possibilities, but the tendency towards increasing concentration is still dominant. The rate of increase in concentration depends on the rate of technological change, the technological regime, and the ease of copying or imitation, among other factors. With rapid technological change and difficulty in imitation, the tendency towards concentration is most pronounced, confirming the arguments of Phillips (1971).

A wide range of other theoretical literature tends to find this persistent dominance, or 'success breeds success' result. One very influential and striking example of this is Dasgupta and Stiglitz (1980a, b). In this model, only one firm engages in R&D and the incumbent is better placed to exploit an innovation than any entrants. More generally, the analysis of patent races finds a strong though not invariable tendency towards persistent dominance. For an excellent introduction to this literature see Tirole (1988).

A different strand of this literature which also predicts persistent dominance is the economic theory of de facto standards, which was discussed in Chapter 7. In a deeper sense, however, it could be argued that the emergence of standards is not a concentrating trend because open standards facilitate entry to markets by removing the proprietary rights around leading designs. Nevertheless, it can be argued that closed (de facto) standards do tend to favour the incumbent.

Similar first-mover advantages apply to producers of pioneering brands, even outside a standards context. For example, Schmalensee (1982) argues that even when competition sets in, pioneering brands have the advantage of appearing less risky than new brands, as the pioneering brands are known while the new brands are not. As one would expect, this issue has been analysed in the marketing literature (see, for example, Urban et al., 1984). Another area of the economics literature, that of product proliferation, suggests that rapid product innovation that takes the form of proliferating slightly differentiated brands will also act as a barrier to entry, and hence will reinforce concentration in a market (Schmalensee, 1978).

One further strand of the literature that finds rapid technology change to reinforce concentration is that which looks at how the cost of R&D increases with speed. This suggests that rapid change is likely to be concentrating because development costs will increase as the speed of development increases (Scherer, 1984: Chapter 4). Wyatt (1986) likens the research process to searching along the branches of a tree. If speed is not a consideration, the firm follows one research line at a time until it finds the best line and the overall number of research lines followed – and hence cost – are minimised. On the other hand, if speed does matter, then the firm will operate several research lines in parallel, and while that strategy will lead to a faster discovery of the successful research line it will also lead to higher costs as more lines are pursued.

It would be wrong, however, to imply that *all* economic theory presumes in favour of persistent dominance. There are parts of the patent race literature (e.g. Reinganum, 1985) in which drastic innovations give the entrant a greater incentive to innovate than the incumbent. Gort and Klepper (1982) suggested that many major innovations would emanate from new entrants, and would tend to occur in the earlier stages of the product life cycle, while many minor incremental innovations would be introduced by existing producers and would occur throughout the life cycle. In a simulation model of innovation in different technological regimes, Winter (1984) found that in the entrepreneurial regime, new entrants were responsible for about twice as many innovations as incumbents, while in the routinised regime, established firms were responsible for the vast majority of innovations.

Organisational Behaviour

While most (but not all) economic theory tends to dwell on the possibility of persistent dominance, a first look at some of the organisation literature suggests a concern with what we shall call organisational inertia. A central idea here is that radical innovations can present severe difficulties for established firms (Daft, 1982; Tushman and Anderson, 1986) and are easier for the small and new firm to exploit.

In fact, rather than the dichotomy, radical versus incremental, Tushman and Anderson (1986) use the slightly different classification of *competence-destroying* and *competence-preserving* innovations. The former occur when mastering the new technology requires fundamentally different competencies from the firm's existing competencies, while the latter can be mastered by firms using their existing technological competencies. Tushman and Anderson argue that competence-destroying innovations are initiated by new firms, while competence-preserving innovations are initiated by existing firms. Tushman and Anderson point out those competence-preserving innovations need not be minor, and indeed can represent 'order of magnitude' improvements in a technology, but the key is that they do not render obsolete those skills that were used to develop the previous generation of technology.

Competence-destroying innovations would appear to have a clear economic implication. We would expect them to lead to greater market turbulence (Audretsch, 1992), and that in the long run they would lead to major changes in industry structure (Chandler, 1962, 1977; Mensch, 1979) and a redistribution of power and control within the firm (Barley, 1986).

The organisation literature has for some time suggested that large incumbent firms are slow to adapt to the challenges of competence-destroying innovations. Thus Burns and Stalker (1961) suggested that organisational structure may reflect such inertia. Burns and Stalker advanced the influential distinction of organic and mechanistic organisations. The latter had a clear organisational structure and were best suited to stable and predictable market conditions, in which they could exploit scale economies. The organic form, conversely, was the best for rapidly changing conditions, especially where new problems and requirements could not be broken down and distributed within an existing mechanistic structure.

Hannan and Freeman (1977, 1984) suggested in more detail a variety of reasons why established firms may exhibit structural inertia. These include internal pressures such as sunk costs, limits on the information received by decision makers, existing standards of procedure and task allocation.

Moreover, they argued that any organisational restructuring disturbs a political equilibrium and this may cause decision makers to delay (or even postpone) reorganisation. There are also external constraints leading to inertia, including exit costs and limits on the availability of external information. For these reasons, Hannan and Freeman argued that the ability of established firms to adapt to a rapidly changing environment was limited. Indeed, this may mean that old organisations become moribund, as they cannot adapt fast enough. McKelvey and Aldrich (1983), among others, advanced an evolutionary perspective on the development of organisational forms in the face of rapid technology and (more generally) environmental change. Child (1984) reviewed the theoretical analysis and practical experience of organic system design.

These ideas have filtered into the economics literature, too. Nelson and Winter (1982, Chapter 5) argued that large complex organisations depend in large measure on tried and tested innovative routines, and are poor at improvising coordinated responses to novel situations. These routines can be seen as a truce in intra-organisational conflict, and Nelson and Winter note that such a truce is valuable and should not be broken lightly. In short, when routines are in place, it is costly and risky to change them radically.

As Mansfield (1968a) notes, the exploitation of synergies depends on inter-divisional communication about diverse experience, perhaps through the exchange of personnel and experience (Aoki, 1986). Cohen and Levinthal (1989, 1990) recognise that this is part of the *absorptive capacity* of the firm. As they point out, absorptive capacity does not just refer to the firm's interface with the external environment, but also to the transfers of knowledge across and within sub-units of the firm. Nelson and Winter (1982) point out that an organisation's absorptive capacity does not depend so much on the individuals as on the network of linkages between individual capabilities. And as Simon (1985) points out, it is learning and problem solving from diverse knowledge bases that is most likely to yield innovation. To keep aware of new technological developments and the competence to deal with them, firms must sustain their multiple technological bases (Dutton and Thomas, 1985).

Finally, we should note some important work in strategic management in the area of dynamic capabilities and competence (Teece et al., 1991). Unlike other analyses of strategic interaction, this recognises that the scope for short-term strategic reorientation is limited. Teece et al. argue that current competence and routines set bounds on the possible technological paths followed by firms, and hence set bounds on the ongoing accumulation of capabilities. This means that radical change poses severe difficulties for large and complex organisations.

CASE STUDY: WILL THE INTERNET LEAD TO PERFECT COMPETITION?[3]

We now turn to what may at first sight seem like a very different question. But in fact it concerns the very same issue. What are the implications of one particular and pervasive innovation on market structure?

Over the last ten years or so, several commentators have claimed that use of the Internet in e-business would help to bring about perfect competition. For example, a survey of Internet economics in *The Economist* (2000) concluded:

> the Internet cuts costs, increases competition and improves the functioning of the price mechanism. It thus moves the economy closer to the textbook model of perfect competition

There are several steps in this argument, but two of the most important are these: (1) that the Internet will undermine monopolistic pricing schemes such as 'noisy' price discrimination; (2) that barriers to entry into e-commerce are low.

This section takes a critical view of these arguments. It is unlikely that the Internet can really stop price discrimination altogether. Some forms of price discrimination are probably 'Internet resistant' and moreover, online retailers can use potential of the Internet to devise ever more subtle pricing schemes. Second, while technological barriers to entry into e-commerce may be low and falling, marketing barriers to entry are substantial and growing. In short there must be doubt that the Internet will actually bring the economy closer to perfect competition.

The End of Price Discrimination?

In many markets (from mobile phones to financial services) the consumer is faced with a bewildering array of different products, services and prices. Economists know that often this is not an accident, but evidence of price discrimination by suppliers. Rather than set a single price for all buyers, the supplier's goal is to extract higher revenues from those prepared to pay premium prices, without losing the custom of those who are not.

Price discrimination generally works along four dimensions: time, space, buyer characteristics and product differentiation. Some of this variation in prices is systematic. Thus, for example, we know that peak-time tickets are more expensive than off-peak. This is unremarkable, and we live with it.

Some of the price variation, however, is 'noisy'. Prices vary over time and from store to store in a random and unsystematic way. Salop (1977) explained the rationale for this. Those consumers with a low opportunity cost to their time can search out the best bargains. Those consumers with a high opportunity cost to their time do not search and expect to pay over the odds. Since the first group tend to be unwilling to pay high prices, while the second group are often prepared to pay high prices, this 'noisy' price discrimination sorts consumers very effectively. Marketers have more recently coined the term, 'confusion marketing' to describe the practice of segmenting markets in this manner.

This 'noisy' price discrimination only works if search is time-consuming and costly. But with the Internet it becomes much easier and less time-consuming to monitor this 'bewildering array' of prices. For example, several web-sites continually monitor all available mortgage products so that the potential borrower can feed in his/her details and then receive an individualised ranking of the best deals. This property also increases the effectiveness of the price mechanism. If a new entrant offers a bargain, then any regularly updated survey can bring this to consumers' attention very rapidly.

Optimists believe therefore that the Internet will undermine price discrimination and thereby force prices down towards perfectly competitive levels. But this argument needs careful scrutiny.

Are Some Sorts of Discrimination 'Internet-Resistant'?

The Internet can reduce the cost of searching to find the best price for each type of product. But if the consumer already knows where the low prices are, then it is not clear that the Internet offers him/her much that is new. For that reason, it is arguable that *systematic* price discrimination is essentially 'Internet-resistant'. If people know that peak-time travel is relatively expensive but still choose to travel at peak-time, then the Internet doesn't change anything.

Some forms of 'noisy' price discrimination, by contrast, are clearly *not* 'Internet-resistant'. 'Noisy' pricing by space is the obvious one. If different stores in different locations charge different prices, then the Internet offers a quick and effective way of searching to find a bargain. Those stores charging over the odds will find that they do little trade. More or less the same argument applies to noisy pricing by product differentiation.

What about noisy price discrimination by time? This is less clear. As any frequent user of the Internet knows, it is a surprisingly timeless place. Web pages indicate today's prices, but with the obvious exception of pages listing stock prices, they don't generally indicate whether today's prices are high or low relative to yesterday's or tomorrow's prices. The regular surfer will be able to tell, but the infrequent surfer may not. So this form of price discrimination may be Internet-resistant for some consumers.

The most Internet-resistant form of noisy price discrimination, however, is by customer characteristics (such as usage intensity). The point here is that *only part* of the 'noise' faced by the consumer relates to prices charged. The rest stems from the fact that the customer may not know in advance how intensively (s)he will use a service. As a result, even with complete information and understanding of all the available tariffs, the consumer does not know which choice to make. The Internet can remove some of the noise – that part which relates to random variation in tariffs – but cannot remove the other part.

This may not be a problem with some online tariffs. For example, consumers soon learn about their pattern of electricity consumption, and are therefore able to work out the best tariff for their pattern of use. But with new products and services, and especially those subject to rapid technological change, the consumer may be far from certain. Many new mobile phone users originally saw themselves as 'light users', and hence chose a tariff offering a low fixed cost but high marginal cost. But many of them have turned out to be heavier users than they expected, and would do better to switch to a different tariff.

What New Forms of Discrimination Does the Internet Enable?

The argument so far has taken a given degree of price dispersion and shown to what degree the Internet gives a consumer the power to get the better of discriminatory schemes. However, that is not the end of the story.

The prevalence of price discrimination suggests that the desire to be profitable is a pretty potent force in most industries. If the Internet confounds a supplier's price discrimination strategy then profit margins or sales (or both) are reduced. So what can the supplier do next? (S)he can seek to create a new dimension of price discrimination.

Swann (1990) showed that product innovations have greatest effect when the characteristics space in which these products are located gets congested. When competition is intense, because the space is congested, then there are strong returns to the innovator who can add extra product characteristics that move his/her product away from the competitive throng. When competition is weak, on the other hand, and the characteristics space is only sparsely populated, then the incentive for product innovation is more limited.

Exactly the same argument applies in price discrimination. The supplier is able to charge different prices only when the products offered differ along one or more dimensions. But the Internet can reduce the competitive distance between products. If the supplier wishes to maintain price discrimination (s)he must recreate this competitive distance and that requires differentiation along another dimension. The incentive to create radical price discriminatory schemes is greatest when the Internet is closing in on existing schemes.

One manifestation of this, especially common in financial services and telecommunications is the frequent realignment of tariffs. These changes will not perhaps confuse the consumer for very long, for eventually the Internet will be able to track all these changes and their implications. Nevertheless, each realignment works for a while.

Some scholars in the field of marketing have suggested that the most persistent 'confusion marketing' strategies can make it impossible for *any* customer to work out the best deals. But from the economist's point of view, that doesn't make sense. The key is that those with a low ability to pay (but a high willingness to search) end up paying lower (expected) prices. If noisy price discrimination is to work, it is essential that *some* people can solve the puzzle of prices, even if many others are confused. It is no good if all consumers are confused, because then no market segmentation is achieved and all pay the same (expected) price.

Even more interesting, the same technology also allows the seller to devise ever more complex and personalised pricing schemes. For repeat-purchasing customers with a long history of purchases, the online supplier can devise a customer-specific tariff that approaches first-degree price discrimination. The effect of Internet upon noisy price discrimination is a bit like the effect of antibiotics on bacteria. We now recognise that antibiotics don't destroy all bacteria permanently. Sometimes the bacteria mutate into forms that are resistant to the antibiotic. Likewise, when the police launch a new crime prevention strategy we don't expect that to get rid of crime for ever. The canny 'crook' can generally devise ever more skilful methods to evade detection.

Low Barriers to Entry?

The second key proposition made by those who believe that the 'New Economy' will be perfectly competitive is that barriers to entry into e-commerce are very low. This argument is partly right but partly wrong.

In one clear respect the argument *is* right. The *technology* costs of creating a web-site, on their own, are no barrier to entry. And once that web-site is up and running, it is accessible by anyone, anywhere. The village store can for a small outlay start to compete in the metropolis – in principle, at least.

However, that argument misses the point. While it is cheap to connect, *successful commercial exploitation* of the Internet involves substantial fixed costs. These are not technology costs but marketing costs.

Connection does not ensure visitors. It is well known that visits to web-sites are highly concentrated in a relatively few sites. While a search engine will in principle list all those trading in a particular sector, in practice most customers will visit the sites that come at the top of the list, the site with the best-known brand name, or the site with the most links from other sites.

For many reasons, there are substantial first-mover advantages in e-commerce. The first movers tend to come at the top of the search engine's lists, because a principle of seniority applies. They tend to have the best 'dot.com' names, because they got in early enough to buy these names while available. The folklore of the dot.com world has defined several guidelines for good names, including the so-called 'Radio Test': the dot.com name should be memorable if heard just once on the radio. In view of this it is hardly surprising that good names are so highly valued: the best 'dot.com' names can trade for six or seven figure sums.

The first movers' web-sites are often the first points (or 'portals') through which users access the Internet. Sites located close to these portals (with links from the portals) enjoy disproportionate competitive success – in just the same way as the well-located concession in a major airport or station. In addition, first movers have had longest to gain the trust of online consumers. Moreover, since many companies personalise their web-sites to individual customers, those customers build up familiarity and expertise in using a particular site. This expertise is partly site-specific, and hence the customer would experience switching costs in moving from a familiar site to a new site. All this helps to lock users into sites with which they have become familiar.

The first movers also enjoy the highest advertising revenues. The reasons for this lie in the basic economics of advertising. Advertisers ideally want to achieve a large reach (cumulative audience) with an optimal frequency of exposure. Far and away the best way to achieve this is to buy access to a relatively small number of large audiences. To buy access to a lot of small audiences is far less effective, for it runs the risk that some of the target audience will not see the advertisement on any occasion while others will be bored to distraction by over-exposure. For this reason, the money value of an audience is a highly non-linear function of audience size and only those sites with large visitor numbers can expect to make much advertising revenue.

Some dot.com companies had to spend huge sums on establishing their brand name. This is a substantial fixed (and largely sunk) cost, and hence a formidable barrier to commercial success. Hutton (2000) believes that well-established brands in the media and elsewhere can enjoy a strong head start on the Internet, though some point out that brands from other sectors do not necessarily work in e-commerce.

All of these arguments suggest that the race for control of the Internet has a 'winner takes all' character to it. In the economic analysis of product and standard races, the first mover enjoys a lasting advantage from being quick to build up a substantial network of users. That generally means that markets with strong network effects become highly concentrated. The same argument applies here.

In our survey of technology and market structure at the start of this chapter, we concluded that much technological change had led to greater concentration. But we also noted some cases where the trend was towards less concentration. The rapidly declining cost of IT has definitely made it possible for small enterprises to compete effectively in some IT-intensive industries. And if that were the key to success in e-commerce, it would be reasonable to expect the Internet to reduce industrial concentration. However, as we have argued above, the barriers to success in e-commerce lie elsewhere.

Implications

There should be no doubt that a technology with such potential can have a huge impact on the nature of competition. Schumpeter (1954) described how, 'competition from the new commodity, the new technology, the new source of supply, the new type of organisation', is so effective that, 'it becomes a matter of comparative indifference whether competition in the ordinary sense functions more or less promptly'. Schumpeter would have been very interested in the Internet.

Some have gone as far as to argue that the Internet will make the economy more like the textbook model of perfect competition. In this last section, we have looked at two steps in that argument: (1) Will the Internet bring an end to monopolistic and discriminatory pricing? (2) Will the Internet reduce barriers to entry into e-commerce? In both cases, the answer seems to be 'no'. Some discriminatory pricing practices may not be Internet-resistant, but many are, and in any case the same technology makes it possible for online suppliers to invent new methods of price discrimination. Moreover, while the technological barriers to entry in e-commerce are small and declining, the marketing barriers to commercial success are large and growing. All this must cast doubt on the idea that the Internet will make the economy more like perfect competition.

The Internet is deceptive. It seems to help the village store compete in the metropolis, but it also helps the superstore compete in the village. In effect, the Internet places the two side by side in cyberspace. And in general, as we all know, if a small grocery store is located next to a supermarket, it is usually the small grocery store that suffers. This is the same sort of paradox that we found in Chapter 13 on clusters.

NOTES

[1] This section is an abbreviated and updated extract from my Chapter 2 in Swann and Gill (1993).

[2] This section is an abbreviated and updated extract from my Chapter 2 in Swann and Gill (1993).

[3] This last section is an abbreviated and updated extract from Swann (2001b).

19. Innovation and wealth creation

My aim in this chapter is to get the student to open his/her mind. Most economic thought is imbued with a rather simple model of how innovation creates wealth and economic growth. We met this in Chapter 2 and will have a closer look at it below. But this simple 'linear' model has been subjected to a great deal of criticism. The most important defect is that it encourages us to think that the impact of innovation on wealth creation is all channelled along one path. That is definitely wrong, and when we use this simplistic model, many other channels through which innovation can impact on wealth creation remain hidden. Indeed, I would venture to say that some of these hidden linkages may, in some circumstances, be much more important than the conventional channel. However, we cannot know that until we start to research these linkages.

After summarising the simple linear model, I shall introduce a more complex interactive model. Then in subsequent sections I shall give a rough sketch of *some* of the many other linkages at work in my more complex interactive model of innovation and wealth creation. In this short chapter, I shall not attempt to survey the evidence on each link, which will be a book in its own right.[1]

Before embarking on that journey, however, I need to take a short digression on the definition of wealth.

WEALTH: MERCANTILE AND RUSKINIAN

It is not easy to define wealth. Nevertheless, as J.S. Mill asserted in the introduction to *Principles of Political Economy* (1848/1923, p. 1): 'every one has a notion, sufficiently correct for common purposes, of what is meant by wealth.'

At one level, it is hard to disagree with that. We know that a wealthy country is capable of producing a large stock of useful things, whether products or services. We know that a wealthy country has a large stock of financial reserves which underpins its international trade and thereby allow the country to procure more useful things. In all this we are, for the most part, talking about *material wealth*: tradable goods and services.

One of the authors we met in Chapter 2, John Ruskin, held a very different opinion, however. He took exception to Mill's remark, and responded: [2]

> There is not one person in ten thousand who has a notion sufficiently correct, even for the commonest purposes, of 'what is meant' by wealth; still less of what wealth everlastingly *is*.

Ruskin believed that this lack of precision about what constitutes wealth was having very unfortunate implications. Industrialisation and laissez-faire appeared to be a path to maximising material wealth. But in Ruskin's view, industrialisation and laissez-faire were certainly *not* the right paths to maximising wealth in the sense that he understood it. Ruskin made an essential distinction between *mercantile* wealth, meaning traded wealth,[3] and wealth in a broader sense. He suggested a definition of beautiful simplicity: [4] 'There is no wealth but life'.

This brief statement gives a perfect summary of what is meant by *Ruskinian wealth*. Ruskinian wealth is much closer to *quality of life*. Ruskin observed that many who were wealthy in a mercantile sense were not capable of real wealth because they did not know how to turn their material wealth into real *quality of life*.

Now, some economists might acknowledge this distinction, but respond that economics is only concerned (and can only be concerned) with mercantile wealth. But I don't accept that. There is much discussion in the recent economics literature on the relationship between economic growth and happiness, and that finds the relationship is by no means straightforward. We cannot divorce our study of wealth creation from the discussion of wealth in this broader sense. In what follows, I shall use a broader *Ruskinian* definition of wealth.

A SIMPLE STORY

Figure 3.1 introduced the first few steps of a simple 'linear' model of innovation. Figure 19.1 illustrates the rest of it. (In fact, we have compressed the four stages of Figure 3.1 down to two stages here.) With luck and hard work, research and creativity will generate promising inventions and after a lot of development and design work these can grow into commercially viable innovations. What happens next? The innovations are adopted in the workplace and after a while the company will be in a position to offer some new and/or more attractively priced products in the marketplace. If these are of interest to consumers, then these products will be bought and consumed. And as a result, the consumer will be better off – both in terms of material wealth and, we hope, in terms of welfare.

Figure 19.1 Innovation and wealth creation: simplistic linear model

According to this simple model, then, the wealth-creating effects all follow one channel. So, the model asserts, creativity and invention can only create wealth if channelled through innovation. And innovation can only create wealth if it is channelled through the workplace and through the outputs of the workplace which are sold in a product market. And the model then asserts that the only route to wealth creation and welfare is through consumption.

It only takes a little reflection to see what an extremely limited viewpoint this is. First, if we give the matter some thought, it is not hard to think of other channels through which creativity may enhance welfare and wealth (in a *Ruskinian* sense). To take just one example, many enlightened people follow hobbies that use their own creativity to enhance their *Ruskinian wealth* and welfare. Second, the viewpoint is far too limited because it neglects *reverse* linkages. In the model, for example, there is no feedback from innovation to creativity, and yet in practice there are (or should be) many such linkages.

This simple linear model also leads us to make other, potentially serious, errors. First, if the only route to wealth creation is through the outputs of the workplace, then it is understandable that we should be preoccupied with productivity growth, and indeed many economists are indeed very preoccupied with that. We saw in the quotation from Solow in Chapter 2 that innovation is indeed an important source of productivity growth. But some, indeed, have gone as far as to suggest that innovation *only* matters in the economy to the extent that it increases productivity. That seems to me a gross error: there is much more to innovation than productivity growth alone.

To avoid such errors it seems to me essential that we set out a much more complex and interactive model of how creativity and innovation create wealth. We do this in the next section.

A COMPLEX STORY

The model developed below (Figure 19.2) could be said to leave no holds barred. In principle, *everything relates to everything else*. This makes it very complex and very multifaceted. But I think it is important for the student to start to learn how to think within such a complex model because that is a much closer approximation to reality than Figure 19.1.

In Figure 19.2 we introduce one further category in addition to those in Figure 19.1. This is the environment. Our reason for doing that is threefold. First, we shall note in Chapter 21 that some innovations, unintentionally perhaps, can have adverse effects on the environment. Second, if we take the concept of Ruskinian wealth seriously, the environment itself is still an essential source of Ruskinian wealth for many people. And third, there is some evidence that a favourable environment can have a beneficial effect on some of the other activities in Figure 19.2.

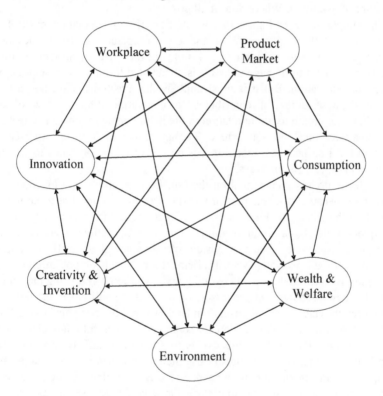

Figure 19.2 Innovation and wealth creation: complex interactive model

Some readers may feel slightly bewildered by Figure 19.2. Are *all* of these linkages really important? And where are we to start? Concerning the first question, I don't assert that all linkages are equally important but I suspect that all of them exist and we would do well to try to understand them. Concerning the second question, we can start anywhere we like, but I shall progress around the diagram in a clockwise direction, starting with creativity and invention. In each of the seven subsequent sections, I shall list *some* of the potential effects of each variable on the others in the diagram. Many of these will be positive relationships but some are negative.

SOME EFFECTS OF CREATIVITY AND INVENTION

Let us suppose that the oval marked 'creativity and invention' in Figure 19.2 represents all the creativity and invention in the UK. That is a lot of creativity and a lot of invention. Where does it all go?

The simple model of Figure 19.1 would have us believe that either (a) it is all channelled through innovation and the workplace; or (b) it does not contribute to wealth creation. I do not dispute that organisations find creativity difficult to manage unless it is channelled through the discipline of design or innovation. But it is most unlikely that creativity plays no part in wealth creation unless it contributes to innovation. Only a few of us contribute our creativity to innovation in the Schumpeterian sense. The rest of that creativity must go somewhere. Perhaps some of it is lost to wealth creation but I suspect that much of it is not. Rather, it contributes to wealth creation in quite different ways.

We have already commented on the link from creativity and invention to innovation because that was part of the simple linear model. Let us continue with the other linkages. First, is there a direct linkage from creativity to the workplace which bypasses innovation? In many organisations, there is. Some innovative companies make a virtue of allowing their staff a certain percentage of the working week (perhaps 10 per cent, or one afternoon a week) to give vent to their own creativity and invention by pursuing their own ideas for new products and processes. Many of these will never see light as commercially viable innovations and indeed, it is not the company's intention that they should – though occasionally some very successful innovations may stem from this. Rather, the objective is to encourage staff to develop their own human capital and in doing so raise their morale and commitment to the company. Some organisations, indeed, use this as a kind of non-pecuniary perk to retain their most capable staff. Or it could be seen as part of an *efficiency wage* strategy. Company A offers this perk while most competitors do not. This means that staff will be motivated to work hard for company A because they know that if they shirk and are dismissed, they will no longer be able to enjoy the perk. In this way, allowing staff to use their creativity within the workplace will enhance productivity in the workplace but not because it contributes to innovation.

In the same way, it seems certain that there is a direct link from creativity to consumption, which bypasses innovation as such. We have already written in Chapter 15 of the active (or innovative) consumer. Let us, once again, remind ourselves of Marshall (and McCulloch's) description of this consumer (already cited in Chapters 2 and 15):

> The gratification of a want or a desire is merely a step to some new pursuit. In every stage of his progress he is destined to contrive and invent, to engage in new undertakings; and when these are accomplished to enter with fresh energy upon others.

This consumer is 'destined to contrive and invent'. And in doing that, (s)he will often use his/her own creativity rather than buy such creativity in a marketplace. This use of the consumer's own creativity is indeed an essential part of wealth creation for the Marshall consumer. We do not increase the welfare of the Marshall consumer just by getting him/her to consume more and more. Rather, the welfare of the Marshall consumer is increased when (s)he uses his/her creativity and some purchases in the marketplace to 'engage in new undertakings'. We noted in Chapter 15 the idea of consumption as household production. From this perspective, we could say that the households use their own creativity to produce *more* from a given bundle of purchased goods and services. While I cannot quantify it, I suspect that this use of creativity may be just as important in wealth creation as that creativity which is channelled through innovation!

Again, there is some linkage from creativity direct to wealth creation. So, for example, some people get pleasure from creative writing even if they know that they will find no market for their work and even, perhaps, that nobody else will ever read their work. Equally, many get pleasure from sketching and drawing even if, once again, it is purely for themselves and they will never sell their work.[5] Now, it could be argued that both these activities require some consumption (pen and paper) so should be seen as creative consumption. But here, the element of consumption is so trivial that I prefer to call this a direct link from creativity to (Ruskinian) wealth creation. Now, of course, it is arguable that there is also a negative relationship here. Creativity can be a painful process (see Chapter 9) and that may lead to poverty and alcoholism, both of which reduce Ruskinian wealth as well as material wealth.

Finally, we can see a direct linkage from creativity to the environment. At a modest level, I make my garden a more pleasant place by using my creativity. This is not pure creativity, perhaps, because there is some outlay on plants and tools. But I would not call this hard work 'consumption'. Others do the same in their own houses: they use their own creativity to make their houses more attractive. It may even become an art form with *feng shui*. All of these applications of creativity contribute to Ruskinian wealth.

SOME EFFECTS OF INNOVATION

Now we turn to the effects of innovation. Many of these effects are directed at and felt in the workplace, for sure. But that is not all.

First, there is, or should be, an important feedback from innovation to creativity and invention. This is something that concerns governments a great deal. The British government is constantly arguing that researchers in universities should improve their dialogue with innovators in companies. Some business-people argue that this is necessary to ensure that academics do 'business relevant' research and are discouraged from doing the 'blue skies' research of no obvious industrial applicability. Other more enlightened business-people recognise it is better if academics keep their 'blue skies' work going, but in doing it take note of what is happening in industry and therefore how industry might realise commercial benefits from the research. Many academics find that dialogue with industry and policy raises many interesting research questions and in my opinion that is an essential feedback mechanism.

Second, we could argue that some innovation is destined directly for the marketplace rather than the workplace as such. By this, I mean innovations such as the supermarket or e-business which shape the marketplace itself rather than change the products and services available in the marketplace. The supermarket has been an immensely powerful – almost unstoppable – retail innovation. It is not so much that the goods and services traded in supermarkets are different from what is available elsewhere, though obviously supermarkets offer very wide choice and in some cases at very low prices. Rather, the power of the innovation is the convenience it offers the consumer. Equally, e-business is an innovation which would have been of huge interest to Schumpeter because it does, in effect, create a new sort of marketplace.

Undoubtedly, the supermarket has played its part in wealth creation, but as an innovation it works in a rather different way from the simple linear model described above. But also, as supermarkets become ever more dominant in the retail scene, we see the negative side of this retail innovation. First, the supermarket has displaced other retail outlets on the high street and that imposes costs on those without cars (especially the elderly). Second, supermarkets are responsible for a substantial carbon footprint, because of the additional car journeys generated by this retail innovation.

There are clearly some very important linkages from innovation to the environment. Some of these are benign. Innovative town planning has revived old industrial cities by making old warehouses and other industrial buildings into attractions in their own right. This is apparent in the docklands of Liverpool, the centre of Manchester, and in the Lace Market district of Nottingham. Equally, innovative landscape gardeners have achieved the same effects as those described above, but in this case in city parks as opposed to private gardens.

We can also expect to find benign linkages from innovation to the environment in the form of clean technologies, or greater fuel efficiency and less noise from cars and aeroplanes.

However, we have to recognise that there are potentially very many negative links from innovation to the environment. Some of these are obvious enough. So, for example, in the industrial revolution, some factory innovations may have achieved greater productivity in the workplace, but they also created air pollution, water pollution and environmental pollution more generally.

Some of these effects are less obvious, however. Let us take one unexpected example. Since the mid-1980s, we have seen a long sequence of innovations in personal computer operating systems (from MS-DOS to various generations of Windows). We might imagine that this is a very clean innovation (it is *software* after all). But we are now starting to realise that these innovations may be responsible for a huge amount of e-waste. How can that be? The point is that each subsequent upgrade in the operating system requires a computer with more processing power and more memory. There comes a point when a computer that is perhaps only six to eight years old is obsolete in the sense that it cannot run current software, though it is still perfectly capable of running old software. Many environmentalists are deeply concerned about the growing trade in e-waste, products that still work but are obsolete in the sense described above, and which are shipped to third world countries to be dumped in landfill. In short, something that at first sight appears to be a clean innovation has some very adverse environmental side-effects.

SOME EFFECTS OF THE WORKPLACE

The simple model of Figure 19.1 recognises that the success of a company depends on how that company succeeds in a marketplace. But there are other linkages from the workplace within Figure 19.2.

There can be a very important feedback from the character of the workplace to creativity and innovation within a company. Some enlightened companies have learnt this to their advantage and other less enlightened companies have learnt it to their cost.

In pioneering work, Ekvall (1987, 1996) has developed the concept of a *'creative climate'* and developed a technique for measuring the creative inclination of a workplace. This identified ten dimensions to creative climate and his questionnaire sought to measure these. This is very important, because some have argued that creative climate or creative culture is the single most important influence on the innovative potential of the company. Zaltman (here quoted from von Stamm, 2003, p. 335) says:

> the daily environment provided by a firm is the single most important determinant of innovative thinking among its personnel. An effective intervention in that environment is far more productive than efforts to intervene in the individual manager's thinking.

The character of the workplace can have obvious effects on the consumption behaviour of its employees. Enlightened employers will be concerned with the health and welfare of their employees and may try to promote healthy consumption. Other less enlightened employers place their employees under so much stress that they eat and drink to excess. Some people who take on highly paid jobs complain that while they earn more their quality of life is no better. One possibility is that busy people face a higher cost of living because, for example, instead of cooking meals for themselves they eat out in expensive restaurants. Another possibility is that the high salary comes at the expense of various forms of Ruskinian wealth – no quality time to spend with the family, for example.

The workplace can clearly have a direct impact on the environment. We have already commented on how the industrial revolution damaged the environment in which many lived and worked. The following extended (and memorably sarcastic) passage by Ruskin illustrates the point very powerfully:[6]

> Last week, I drove from Rochdale to Bolton Abbey ... Naturally, the valley has been one of the most beautiful in the Lancashire hills ... (but) at this time there are not ... more than a thousand yards of road to be traversed anywhere, without passing a furnace or mill.
>
> Now, is that the kind of thing you want to come to everywhere? Because, if it be, and you tell me so distinctly, I think I can make several suggestions to-night, and could make more if you give me time, which would materially advance your object. The extent of our operations at present is more or less limited by the extent of coal and iron-stone, but we have not yet learned to make proper use of our clay. Over the greater part of England, south of the manufacturing districts, there are magnificent beds of various kinds of useful clay; and I believe that it would not be difficult to point out modes of employing it which might enable us to turn nearly the whole of the south of England into a brick-field, as we have already turned

nearly the whole of the north into a coal-pit. I say 'nearly' the whole, because, as you are doubtless aware, there are considerable districts in the south composed of chalk, renowned up to the present time for their downs and mutton. But, I think, by examining carefully into the conceivable uses of chalk, we might discover a quite feasible probability of turning all the chalk districts into a limekiln, as we turn the clay districts into a brick-field. There would then remain nothing but the mountain districts to be dealt with; but, as we have not yet ascertained all the uses of clay and chalk, still less have we ascertained those of stone; and I think, by draining the useless inlets of the Cumberland, Welsh, and Scotch lakes, and turning them, with their rivers, into navigable reservoirs and canals, there would be no difficulty in working the whole of our mountain districts as a gigantic quarry of slate and granite, from which all the rest of the world might be supplied with roofing and building stone.

Is this, then, what you want? You are going straight at it at present …

But the effects need not be negative. Tourists who have visited Port Sunlight on Merseyside will see what enlightened employers (Lever Brothers) could do for the environment in which their employees worked, and hence for their Ruskinian wealth.[7] The University of Nottingham has turned an old derelict industrial site (the old Raleigh bicycle factory) into a beautiful new campus,[8] which I find is a very pleasant environment in which to work. Moreover, just outside Nottingham we find the delightful Attenborough Nature Reserve. This used to be a collection of gravel pits, and not an especially attractive workplace perhaps, but it has now (with some imagination) been turned into a reserve with an exceptional diversity of wildlife. Indeed, it is arguable that the site would not be so special now had it not been an industrial site before.

EFFECTS OF THE MARKETPLACE

For the economic consumers of Chapter 15, the role of the marketplace is purely instrumental. To them, it is a place in which they buy the goods which they will later consume. They take no particular pleasure in visiting the marketplace: indeed, ideally they would like the whole business of shopping to be done as quickly as possible. But not everyone is like that.[9] Some people find the marketplace a source of pleasure even if they do not buy anything. This includes people who delight in spending hours looking around expensive designer shops and department stores, but it also includes people who delight in visiting much cheaper street markets or 'flea markets'. We could argue that for all these people, visiting the marketplace can create Ruskinian wealth, even if nothing is bought or consumed.

More generally, we can see some sort of linkage from the marketplace to many other categories in our model. There can also be an important feedback from the marketplace to innovation and the workplace. Those visiting the marketplace in a professional capacity can learn much about the state of the market and the nature of consumer demand that will be of value in guiding their innovation strategy and, perhaps, how the workplace is organised. The marketplace can create a pleasant environment or have a less benign effect on the environment. Examples of the former could be the magnificent market squares of some old market towns; examples of the latter have been discussed above in the context of supermarkets. Finally, we note that the marketplace has been a source of inspiration and creativity for many centuries. We can see this is elegant paintings of country markets and in the more idiosyncratic depiction of markets in Salford by L.S. Lowry.

EFFECTS OF THE CONSUMER

The consumer is sovereign. Or so it is sometimes said. A sovereign can surely leave his/her footprint on the economy. Now, not all the consumers we met in Chapter 15 behave like sovereigns. But some do have a marked influence within Figure 19.2.

The work of von Hippel (1988) has documented how influential the consumer (or at least the industrial buyer) can be in helping to guide companies' product innovation strategies. Indeed, innovation surveys, such as the Community Innovation Survey in the UK, have documented how contact with customers is one of the single most important sources of ideas for innovative firms. A well-known example of this is found in the business career of the great entrepreneur, Sir Richard Branson. At an early stage in the history of his Virgin record stores he spent much time talking to teenagers about their record-buying behaviour and used what he learnt from these discussions to create a very successful chain of retail stores. Another well-known example is the phenomenon of 'texting'. Phone manufacturers and operators added this function to mobile phones as something of an afterthought. It was not expected that it would be widely used. But the unexpected consumption behaviour of teenagers, who used text messages far more than voice calls, demonstrated to phone companies that they must take the potential of the text very seriously.

More recent work by von Hippel (2005) goes further than that, however. The customer is not just an invaluable source of information to the innovative company. The customer may actually become *the innovator*. As von Hippel says (recall the quotation is Chapter 2), users are increasingly able to innovate for themselves, and user-centred innovations have many advantages over manufacturer-centred innovations. This would have been seen as a very radical idea ten years ago, when innovation was very much the preserve of the innovative company. But now, especially in an Internet era, it just seems common sense. I see the arguments in von Hippel as very supportive of the framework opened up in this chapter. The linear model of Figure 19.1 is way off the mark because it misses so many important influences on (and effects of) innovation.

Some consumption inevitably has an impact on the environment, sometimes benign, but often not. Some of the onus for reducing the carbon footprint from economic activities lies with consumers. We can insulate our homes, we can use public transport or bicycles rather than drive our cars, we can take holidays by train rather than using the plane, we can make sure not to leave our computers or video recorders on standby and we can do more recycling. However, it is arguable producers have a greater responsibility for ensuring sustainability.

EFFECTS OF WEALTH

In Chapter 15 we met the Veblen consumer. This was – in Veblen's original work – someone so wealthy (in a material sense) that they wished to demonstrate that wealth by conspicuous consumption. Such very wealthy people have left their mark all over Figure 19.2. Here are some brief examples.

The wealthy may leave their mark on the landscape or environment more generally, by building and maintaining fine estates. Some go further and leave their estates to become the location of a university or other place of learning – and thus contribute to creativity and invention. We are very fortunate in Nottingham that the beautiful former estate of Sir Jesse Boot (founder of the Boots Company) is now our main university campus. A more famous example is the great Stanford University in California: this was founded by railroad magnate and California Governor Leland Stanford and his wife, Jane Stanford, and named in memory of their only son who died at the age of 15.

Wealth can contribute directly to innovation. Some wealthy people, sometimes called business angels or serial entrepreneurs, have played a very important role in supporting the innovative efforts of startup companies.

We have discussed already the decision of enlightened employers to create a pleasant workplace for their workforce. This is not just altruism. It is also based around an expectation that a pleasant working environment will help to attract and retain excellent staff who will repay this by making an exceptional contribution to the fortunes of the company.

EFFECTS OF THE ENVIRONMENT

We have already referred to the environment at several points in this chapter. Our original purpose in including the environment in Figure 19.2 was that so many categories in our model could impact on the environment, and given the urgency of environmental concerns now facing us, the economist has to have a framework for understanding whether innovation can help the environment or, in fact, whether innovation makes things worse.

However, the environment itself will have several other important effects within Figure 19.2. The most obvious, perhaps, is the role of the environment in creating Ruskinian wealth. To those who take pleasure in walking in the countryside or visiting the seaside, this idea needs no explanation. Indeed, for some retired people who are still fit enough to do a lot of walking, this Ruskinian wealth may as important as any other source of wealth.

The effects of the environment can be found in other places in Figure 19.2. Some of these have been discussed already. Moreover, just as the creative climate within a company is an important factor influencing creativity and innovation, so too may be the physical environment. Some have suggested that in addition to all the other advantages which make California such an economic success story, the fact that it is such a pleasant environment and climate (meteorologically speaking) makes it somewhere people wish to come to live and work.

In this short chapter we have only been able to scratch the surface of all the practical linkages corresponding to Figure 19.2. The interested reader who wants to explore this further is referred to Swann (2010, forthcoming).

NOTES

[1]   In ongoing work, Swann (2010, forthcoming), I am putting together evidence on all these 'hidden' linkages.

[2]   Ruskin (1905/1996d, pp. 131-132).

[3]   Literally, wealth traded by merchants.

[4]   Ruskin (1905/1996c, p. 105).

[5]   Some great artists might be put in this category. The great Vincent van Gogh only ever sold *one* painting in his lifetime.

[6]   Ruskin (1905/1996b, pp. 336-337).

[7]   Another example is what the Cadbury family did for their employees in Bourneville (Birmingham).

[8]   The Jubilee Campus, where the Nottingham University Business School is based.

[9]   Mumford (1961) reminds us that marketplace in ancient societies was a meeting place where people met for many purposes and *not just* for the exchange of produce.

20. Innovation and competitiveness

We have seen in the previous chapter how innovation can contribute to wealth creation. We argued that the effects on innovation were not limited to the channel described in the simple linear model, but were much more diverse. Moreover, the model we used, Figure 19.2, contained a large number of feedback effects. So, in short, the effects of innovation on wealth creation could follow quite a variety of complex paths.

Now, having said all that, the immediate *reason* why most companies innovate is not, in the first instance, because that increases wealth in the economy. More often, the immediate *reason* is that innovation enhances their competitiveness and is perhaps essential for their survival.[1]

In short, if we focus on producer-centred innovation, we must recognise a potential divergence of objectives. As economists, our objective in encouraging innovation stems from its wealth-creating effects. But those who innovate have a different objective: to ensure their competitiveness and survival. Does this difference in objectives matter?

This is a very fundamental question. If we believe the simple linear model in Figure 19.1, then it could be argued that the difference in objectives does not matter. For in that simple model, the only way in which innovation can impact on wealth is if companies market new and improved products and services or offer better value for money. Companies have their own motivation, albeit a different one, for doing this. And when they have innovated, we can wait for the wealth-creating effects to follow. So even though there is a difference in objectives, innovation still takes place and will succeed in satisfying both objectives.

That argument is too simplistic however. We shall see in this chapter that the value of innovations to the innovator is not necessarily directly related to the value of these to the customer for the innovations. This means that the divergence in objectives may lead to an imbalance in the sorts of innovations we see.

Moreover, when we admit a more complex model like Figure 19.2, then it is much harder to maintain that a difference in objectives does not matter. In Figure 19.2, innovation can impact on wealth creation in different ways. The approach to innovation that might maximise wealth creation may look very different to the strategy for maximising competitiveness of the innovator.

The reader may be surprised that this chapter has been kept so short. But the reason for that is that most of the essential ground has been covered earlier in the book. My objectives in this chapter are threefold. First, I want to have another look at what we mean when we say that innovation enhances competitiveness. Then I shall discuss my assertion that the value of innovations to the innovator is not necessarily directly related to the value of these to the customer for the innovations. Finally, I shall show that to have a preoccupation with the effects of innovation on competitiveness is equivalent to ignoring some important parts of Figure 19.2.

HOW INNOVATION ENHANCES COMPETITIVENESS

We saw, in Chapter 5, a very simple diagram (Figure 5.5) which describes how innovation enhances competitiveness. This is reproduced below.

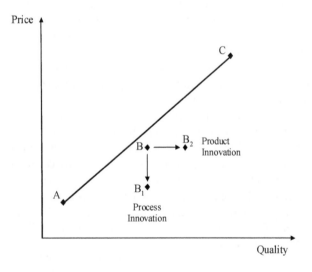

Figure 20.1 Product and process innovation

The diagram shows a product market with three competing products: A, B, and C. As drawn, and if we assume that all consumers have WTP (willingness-to-pay) lines as shown, product B does not look very competitive. Most consumers would prefer A or C, and only a very few (such as those with the WTP line as drawn) will wish to choose B.

But the producer of B can change that if (s)he uses a product or cost-reducing process innovation. The cost-saving process innovation would allow that producer to relocate B to a reduced price (B_1). Alternatively, the product innovation (with no addition to costs) would allow that producer to relocate B to a higher quality (B_2). Both of these moves make B more competitive.

We saw in Chapter 5 that there are some subtle differences in the effects of these two innovations. The student may wish to re-read the section on pages 53-54. Moreover, we should add that the same diagram could equally well be used to represent other dimensions of competitiveness such as: delivery times, the service element, or any other factors that might make a customer choose product B rather than A or C.

THE VALUE OF AN INNOVATION: INNOVATOR AND CONSUMER PERSPECTIVES

The value of the innovation to the innovator is the effect on competitiveness. Increased competitiveness will show up as increased market share (refer back to Chapter 5). Now what we find in such cases is that if a product is only just competitive (that is, it would only be bought by a tiny proportion of customers, with WTP lines as shown), then even a small innovation in B will be of considerable value to the producer. Even a small innovation may be enough to secure a substantial gain in market share. This observation suggests that when the objective of innovation is to steal market share off rivals, then trivial innovations may be far more valuable to the producer than to the customer. A technical analysis of this point is beyond the scope of this book, but the following examples illustrate the point well.

Consider the following selling innovations:

1) A telephone sales team 'cold calls'[2] potential customers to try to persuade them to switch their electricity supplier.
2) Representatives of credit card companies try to persuade shoppers in supermarkets and service stations to take out a new credit card with preferential terms.
3) Companies send 'junk mail' to try to persuade customers to buy an improved product or service.

In each case, these innovations can be very successful from the point of view of the seller because even if only a few customers 'bite' at the offer, this will be enough to make a mark on market share. But the value of these innovations to the customer is very limited – and in some cases is negative! The view that any innovations that are good for the competitiveness of the innovator will automatically be equally good for the wealth of the consumer is too simplistic.

A SUBSET OF FIGURE 19.2

The final point to make here is this: a preoccupation with the effects of innovation on competitiveness will cause us to overlook some of the essential linkages in Figure 19.2. The literature on how innovation enhances competitiveness tends to focus on the linkages identified in Figure 20.2 below. This is only a subset of what we have seen at work in Figure 19.2. Some linkages are ignored altogether; others are treated as if they operate in one direction only. Specifically, our interest in this context will focus on those linkages that influence how the outputs of the workplace will look in the product market – for, in essence, that is what competitiveness is about.

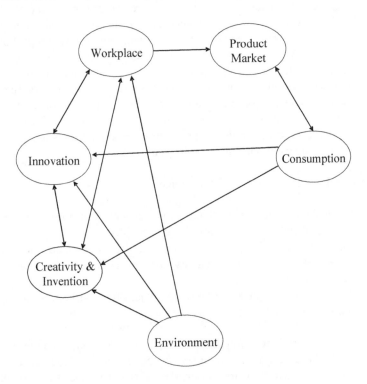

Figure 20.2 Innovation and competitiveness

This view of innovation recognises more than the simple linear model of Figure 19.1. It recognises a role for the customer in influencing the firm's strategy for creativity and innovation, and influencing those who design the marketplace. It recognises that creativity and the environment may have a direct impact on the workplace (as described in the last chapter). And it recognises the potential importance of feedback from innovation to creativity. But it misses everything else.

Does that matter? For those charged with ensuring the competitiveness of a company, *no*, it doesn't matter. They are right to limit their attention to those relationships in Figure 20.2. But for those trying to design policies to promote the wealth-creating effects of creativity and innovation, then *yes*, it *does* matter. The approach to policy that will maximise the effects on competitiveness in Figure 20.2 will not necessarily be the same approach as that which would maximise wealth creation in Figure 19.2. This is an *essential* point, but one that is often overlooked in the business and policy communities.

NOTES

[1] Some exceptional businesses are so successful that they can aim higher than survival and set themselves more ambitious objectives. Some entrepreneurs have said that their purpose is not to make money *per se*. Their purpose is to meet a need in society and if that is done well, then they will also make a lot of money.

[2] 'Cold calling' means when a telephone seller calls a telephone subscriber at random and tries to sell them a product or service which they have not asked for.

21. Innovation and sustainability

Several chapters in this book have been quite short but as I have said before that does not imply the material in them is unimportant. Rather it is because either: (a) many of the required topics have been covered earlier in the book; (b) the required material is covered very well in existing textbooks; or (c) the material, though important, is a bit tangential to the core economics of innovation.

The present chapter will be another short one and once again, that certainly *does not* imply that the topic is unimportant. On the contrary, what topic in innovation could be more important than the issue of how it is harnessed to ensure economic sustainability? In this case, the chapter is short because it is difficult to say much about this topic at this very introductory level. To grasp the full complexity of the relationships between innovation and sustainability would require a much deeper background in economics and innovation than can be assumed for the reader of this book.

However, I urge the student to read this brief sketch very carefully, because it highlights two very important issues that every student of innovation should think about. First, will markets give enough support to innovations that improve sustainability? Second, why does innovation have unexpected side-effects that threaten sustainability?

DO MARKETS SUPPORT SUSTAINABLE INNOVATIONS?

Suppose an innovator comes up with an improved version of a product (X_2) which is more environmentally friendly than the original version (X_1). Will the market provide enough incentive for the innovator to introduce the environmentally friendly version?

A typical economist's response might be as follows. If the environmentally friendly version (X_2) is more expensive than the original (X_1) then the market may not support it. The customer faced with a choice between X_1 and X_2 may stick with former because it is cheaper and the benefits of the latter are not fully understood. Or, even if the customer understands the environmental benefits in principle, (s)he reasons that his/her personal consumption behaviour could only make a negligible effect on the environment. Either way, the market does not support the new version (X_2).

This is a standard argument about the effects of externalities – see Chapter 22 for more detail. If a product (X_1) generates negative externalities for the environment, then there will be a tendency to produce and consume too much for the good of the environment. But if a product generates positive externalities for the environment, then there will be a tendency to produce and consume too little for the good of the environment. And that is what happens here.

How do we resolve this problem of externalities? We shall see in Chapter 22 that the generic solution is to internalise the externalities. This could be done by taxing those who use 'dirty' technologies in proportion to the environmental damage they do, and/or subsidising those who use 'clean' technologies. That is the sort of reasoning that underpins the current approach to carbon pricing: those who leave a large carbon footprint have to pay for the privilege by buying carbon emission permits, while those who think they could adopt a cleaner technology which leaves a modest carbon footprint can benefit from their virtue by selling their carbon permits. This approach is discussed prominently in the Stern Review on the Economics of Climate Change (2006).

Another approach to dealing with the problem of externalities is to set standards and regulations for polluting emissions. By this approach, producers and users simply must adopt sufficiently clean technologies: to continue to use a 'dirty' technology and pay for the right to do so is not an option. In that case, if the new product (X_2) meets the standards and regulations while the old product (X_1) does not, then the market will support the adoption of X_2.

Opinions vary on the relative merits of these two approaches. The relative merits of the two are a little bit like the relative merits of rationing by price and rationing by queuing. If we ration a scarce resource by price, then the rich customers who can afford a high price tend to get what they need, while the poor customers do not. If we ration by queuing, the distributional effects are perhaps fairer, but arguably the effects on economic efficiency are not so favourable because those to whom the rationed good is most valuable get no more than the rest.

CAN INNOVATIONS THREATEN SUSTAINABILITY?

The Stern Review (2006) suggests that innovation driven by an appropriately designed policy will provide the solution for problems of sustainability. But we must also recognise that, while this may not have been what we expected, innovation is ultimately the cause of some environmental damage. Why is that? We can look back over the various chapters of this book to get some clues on this:

- We have seen that innovation can lead to a greater geographical dispersion of the production of components and as a result finished products (e.g. the personal computer) will comprise a much increased number of 'component miles' (Chapters 13 and 14).
- We have seen that innovation can lead to greater intra-industry trade, with similar effects on the total demand for freight miles and passenger miles by air (Chapter 17).
- We have seen that rapid innovation can shorten product life cycles and hasten obsolescence. We mentioned that continuing innovations in computer operating systems had led to the premature scrapping of computers that still work but cannot run the latest software (Chapter 19).
- We have seen that innovation can lead to greater industrial concentration within a small number of clusters. That can have adverse environmental effects from congestion (Chapter 13).

In each of these examples, the adverse effects of innovation were unexpected. And, arguably, the adverse effects are not intrinsic to the innovations themselves. Rather, they arise because the use of these innovations in competitive markets encourages several processes that have some advantages in terms of economic efficiency, but also a significant environmental cost in terms of transport, product obsolescence and congestion.

In short, the original intention of an innovation and its initial effects may appear beneficial (or at least neutral) from an environmental view. But when we have factored in the implications of that innovation for industrial structure, for transport, obsolescence and congestion, the net environmental effects may look very different indeed.

PART VI

Innovation and government

22. Innovation and policy

The subject of policy towards innovation is a huge one and all we can do in this chapter is give a brief introduction. We shall start by trying to answer the question 'why have an innovation policy?' We shall see that part of the answer is to do with market failure. For the sake of students who wish to have a quick refresher, we give a brief summary of what is meant by market failure. We then describe which features of the innovation process may lead to market failure and why. We then summarise the three generic approaches to correcting these market failures, and these three approaches underpin most approaches to innovation policy. The next three sections give a little more detail on each of these generic approaches to innovation policy. The final section offers a postscript on why innovation policy in future may be different from how it is today.

MARKET FAILURE

This section gives a brief summary of what is meant by market failure, why it happens and what in general can be done to correct such failure.

The Concept of Market Failure

While the economist's idea of perfect competition is something of an abstraction in that it can rarely be found, it is of interest as a limiting case because under some conditions it embodies certain desirable properties. First, in perfectly competitive markets, all willing to pay the marginal cost of production will be able to buy a product, which is equitable and efficient. Second, in a perfectly competitive market, the right things are produced and sold, in the sense that products or services valued at or above their cost of production will be produced, while products or services valued below their cost of production will not. Third, a perfectly competitive market is Pareto efficient – meaning that any rearrangement of production activities can only make one consumer better off at the expense of another.

There are, however, certain circumstances in which this rosy picture does not apply. In particular, in these circumstances, the desirable properties described are not obtained. We refer to this as market failure. When there is market failure, some profitable activities do not take place, and/or some unprofitable activities do take place. Moreover, it is not necessarily true that the right things are being produced, nor will the free market outcome be Pareto efficient.

There are three generic reasons why market failure occurs: economies of scale, asymmetric information, and externalities. We describe these in turn, and give some examples of how these sorts of market failure can be overcome.

Economies of Scale

At first sight it may seem a bit perverse to describe economies of scale as a 'failure'. Scale economies allow consumers to buy products at much cheaper price than would be found if there were no scale economies, and hence are a sign of economic success rather than failure.

Nevertheless, the economist would still argue that economies of scale mean that competitive markets fail to perform in the desirable ways set out above. The main reason for this is that if there are economies of scale, arising for example because there are fixed costs of production, then it is not profitable to sell products at marginal cost. The fixed costs must be recovered, and prices set must rise to reflect that. This means that some consumers, at least, who are willing to pay at or above marginal cost, are priced out of the market.

The second reason why economies of scale lead to market failure is simply that where there are continuing economies of scale, the large scale producer will always be able to undercut a smaller scale producer. This is a force towards monopolisation of the market, because ultimately a monopolist producer will have lower average cost than any smaller scale competitor. And we know from our discussion of market structure that monopolists will often try, if they can, to raise prices some way above minimum average cost. Monopoly is in itself a market failure.

The traditional solution to market failure caused by scale economies is not of course to get rid of the scale economies – which, as we said, are in themselves quite desirable. Rather, the solution is to recognise which markets are *natural monopolies*, and to allow such monopolies to emerge, but to place them in the public sector with non-profit maximising objectives, or to regulate them.

Asymmetric Information

This second generic source of market failure occurs in quite a wide variety of settings, but here we shall concentrate on one well-known example, from Akerlof (1970). Take the case of the second-hand car market, where each seller is well informed about the quality (and reliability) of the car they are selling, while any one buyer does not know for certain which are the good cars and which are the bad (or the 'lemons' as they are called in the USA). The buyer may know, roughly speaking, what is the probability that any particular car is good, but (s)he does not know for certain whether a particular car is reliable or not.

This asymmetric information means that the buyer faces a risk that the seller does not. It also creates a problem for the seller of a good car: unless (s)he can demonstrate that his/her car is good, there is no obvious reason why the buyer will be prepared to pay a price premium for that car. Indeed, if buyers cannot distinguish good from bad, then both good and bad cars will sell at the same price – which will be some sort of average of the 'right' price for good cars and the 'right' price for bad cars. Without any method of certifying that his/her car is good, moreover, the seller of good cars may withdraw his/her car from the market: (s)he will consider this market price to be unacceptable.

If sellers of good cars withdraw from the market in this way, then the average quality will decline, and so will the market price. That makes it even more unattractive for the owners of good cars to try and sell their car in this market, and so even more withdraw from the market. Ultimately, we get a repetition of Gresham's Law: bad drives out good.

The market has failed here because the price mechanism has failed to attach a higher price to the good quality car than to the bad quality car. And this failure means that good quality cars may not be traded at all.

There are of course various mechanisms to correct for this problem. Reputable car sellers invest in a brand name for reliability, and may back that up by guarantees. There are independent agencies who will (for a fee) give an informed assessment of the value of a second car. More generally, developed economies have a system of standards and certification, which allows manufacturers to demonstrate that their products conform to certain standards, and are therefore worthy of a price premium. All these devices help to ensure that price and quality are more closely connected, thus removing some of the information asymmetries. This reduces the risk to buyers, but also makes it possible for the seller to get a fair price for a high quality product, and hence corrects (in part at least) the original market failure.

Externalities

The third source of potential market failure is the externality. If A carries out some activity which has a material effect on B, but B is neither compensated for nor charged for this, then we say that A's activity has generated an externality for B. This externality can be negative for example when A's activity generates pollution of some sort, which is a loss of amenity for B. Or it could be positive – for example when A's gardening activity generates a pleasant environment for B, but which B enjoys for nothing.

In either case, there can be market failure. If externalities are negative, then markets will make certain activities look privately profitable when they are socially costly. If externalities are positive, then they can make certain activities look privately unprofitable when they are in fact socially desirable. Either way, there is market failure either because the market will permit some 'wrong' activities to take place, or because the market prevents some 'right' activities from taking place.

Corrections for this sort of market failure tend to be of one of three sorts. First, socially desirable but privately unprofitable activities are run in the public sector (e.g. postal services to remote parts of the country). Second, the government may subsidise activities that generate significant positive externalities, and control or tax activities with negative externalities (e.g. pollution). Third, the provider of positive externalities may try to define property rights over these externalities and charge the beneficiaries a royalty, while those who suffer from negative externalities may sue the producer of these externalities for loss of amenity. Each of these is an attempt to internalise the externalities, and hence correct for market failure.

FEATURES OF INNOVATION ACTIVITIES THAT MAY LEAD TO MARKET FAILURE

The following are some important features of certain aspects of R&D and some other aspects of innovation activity and information exchange which are claimed to distinguish them from the textbook commodity. These features can lead to some kind of market failure, or sub-optimality in market allocation. The market outcome may be sub-optimal in two ways: first there may be *too little* (or conceivably too much) R&D or innovation; second there may be the *wrong mix* of R&D projects or innovation activities.

Economies of Scale – Increasing Returns

It is argued that there are significant economies of scale in certain aspects of R&D: in particular in the production and collection of information. Here we are talking of information in a very pure sense: 'pearls of wisdom', with a high value, at least to some users, which can be imparted in a few words and quickly assimilated.

Information, interpreted in this pure sense, is generally taken to have one of the classic public good properties. The cost of producing information by R&D is predominantly a fixed cost: information can be reproduced at very low marginal cost (in the sense that photocopying a report or copying computer files can be done at very low marginal cost).

As we know from basic first year economics, when there are strong economies of scale in the production of a good, then there is a strong tendency towards monopoly in the supply of that good. Indeed, some have argued that a wide variety of information activities have an element of natural monopoly to them. For example, Metcalfe (1986, p. 44) argues:

> A greater demand [for information] spreads the fixed costs of creation over a greater information output, creates increasing returns to application, and generates a natural tendency towards a monopoly of the information-creating activities in question. In the market for databases, for example, this raises serious regulatory issues which involve not only questions of competition and monopoly in the economic sense but also questions [about] monopoly control over access to information. Similar issues surround the economics of communications networks which also fall into the category of natural monopoly.

The fact that information can be reproduced at practically zero marginal cost means that perfectly competitive markets for the provision of information will fail since competitive firms could never cover their costs in such circumstances. This tendency towards monopoly will probably lead to a tendency towards monopoly pricing and all the attendant inefficiencies.

There is another reason for scale economies in information collection. If information gathering activities are to be worthwhile, they should be done on a large scale. Citing research by Radner and Stiglitz, Dasgupta (1987, p. 9) points out that:

> under a wide class of plausible situations regarding one's attitude to risk there are increasing returns to scale in the value of information at very low information levels. By this one means that it is not worth investing even a tiny amount to learn a tiny bit. If it is worth seeking information of a certain kind, it is worth seeking it in largish chunks.

It might be argued that if information can be reproduced at low marginal costs, then diffusion is not such a serious problem for technology policy, and for information purely interpreted as above, this is probably true; slow diffusion suggests to some extent that transmission costs are significant. It should be noted of course that in many practical cases the assimilation of information cannot necessarily be achieved at zero marginal cost and in some cases assimilation costs can be substantial. Pavitt (1987, and elsewhere) has argued that technology is not costlessly reproducible, that it is locally specific, and that it is not costless to assimilate.

Indeed, to some extent, this idea of R&D as production of information is misleading. Some of the knowledge produced by R&D is tacit rather than codified and cannot be expressed as simple bits of information. Moreover, while information (once produced or collected) may be duplicated at low marginal cost, this is not to deny the scope and necessity for adding value by 'repackaging' information. For many users, *less is more*: the value of information depends on concise presentation, and so there is scope for adding value by customising information in a way that facilitates assimilation. While pure information provision has aspects of natural monopoly about it, selling packaged information is more like the sale of a conventional product. So while there may be market failure in the former, this need not apply to the latter.

Positive Externalities

Following the classic work of Arrow (1962) in the context of perfectly competitive markets, it is often argued that certain aspects of R&D and information production and collection are subject to significant positive externalities. That is to say, one firm's R&D has positive benefits for other firms with related spheres of business. In fact, the spheres of business need not be that related: collaborative R&D may benefit companies with very different product ranges. When such externalities arise, the innovating firm is not able to appropriate the full social benefits of its innovations: in other words the private benefits of R&D or information collection are less than the social benefits. This may act as a constraint on socially valuable R&D programmes. Even though the combined social benefit (to the innovator *and* those who benefit from externalities) exceeds the cost of R&D, the innovator does not recover enough benefit to cover his/her R&D costs, and so cuts back on his/her R&D.

We use the term *informational externalities* describe the information a firm yields to others by undertaking a particular activity. One example would be a firm's R&D programme in a particular area: the very fact that the firm is conducting R&D in that area can convey information about technical feasibility or potential profitability. Or, one firm's R&D or information collection may spill over to rivals as a consequence of personnel mobility, and the fact that information is embodied in products and services produced by that firm.

Again, we note that if one firm's R&D creates benefits for neighbouring firms, then it could be argued that for that group as a whole there is no *problem of diffusion* because the information spills out all too readily. Rather, the problem of market failure is that the company paying for the R&D may not recover enough financial benefit to justify the R&D expenditure – even though the R&D is socially beneficial.

These observations highlight a dilemma in the economics of innovation policy that continually surfaces when we are assessing the policy options. The fact that information can be reproduced at low marginal cost is in a sense a good thing, since it means that many can benefit from an investment; but as we see this property can pose problems for market allocations. If information is 'commoditised', that is individual agents have property rights over it and can trade in it, this removes the problems for market allocation in one sense, but means that fewer agents benefit from an investment in information gathering. And, in general, it seems undesirable to commodify and charge for something that would naturally flow around at zero marginal cost.

While such positive externalities can undoubtedly be expected to exist, it is not clear empirically how important they are. There is relatively little economic research on the identification and measurement of externalities in R&D and information gathering. And it should be noted, of course, that firms in markets can institute joint ventures and 'clubs' to internalise some of the mutual externalities from their separate R&D programmes, and develop property right schemes to improve appropriability. Of course, while this is an understandable strategy in the face of spillovers, it may also encounter resistance from governments concerned with anti-competitive practices; and again, there can be a tension between policies to promote R&D and competition policy.

Negative Externalities

By examining competitive R&D in an oligopolistic context, Dasgupta and Stiglitz (1980a, b), amongst others, have shown that showed that one firm's R&D could yield negative externalities for rivals – by reducing their market shares, for example. This is particularly relevant in the analysis of patent races, for example, where the *winner takes all*. This is sometimes called a common pool problem. A similar result would be found with innovations in a context of minimal product differentiation and discerning consumers: one firm's product innovation can leverage market share off the rival.

In such cases, the private benefit of R&D expenditure to the innovating firm can exceed the social benefit: the innovator benefits by increasing market share while other firms lose. And, although the consumer or user can benefit from such innovation, these increased consumer benefits are not as great as the private gains to the innovator. Loosely speaking this is a result of duplication of R&D effort, which is socially wasteful while privately profitable. More specifically, the losses to the now uncompetitive firm exceed the benefits to the consumer. Net negative externalities of this sort will be most important in the context of oligopolistic competition between firms producing very similar products. They will not be especially important when one firm's R&D activity relates to a product that is not produced by any others.

It is difficult to know whether this outcome, though theoretically possible, is really important in practice. Some writers, for example Stoneman (1987, p. 209), have expressed doubt that the 'common pool' problem has seriously led to excessive privately funded R&D in Britain.

It is of course possible in a particular market structure that positive and negative externalities might be found simultaneously, and to some degree they may offset each other – though it is unlikely that they would *exactly* cancel each other out. In fact a more likely outcome is a bias in the mix of R&D and information collection activities undertaken. In particular, we might find that there is excessive similarity in rival firms' R&D programmes (Dasgupta and Maskin, 1987).

A simple example will show how this might happen. Suppose firm A makes two products X and Y, while firm B makes products X' and Z, where X and X' are very similar products but Y and Z are very different products. Then there will be a bias in A and B's R&D activity towards research related to X and X', where the negative externalities are relatively important (see previous section) and away from Y and Z, where they are not. This is a recurrent theme in the economic analysis of product competition: firms tend to over-invest in innovation when product competition with rivals is intense and under-invest when it is not.

It should be noted that these arguments apply to explicitly oligopolistic markets, and not to perfectly competitive ones. It is also important to note that arguments here suggesting excessive duplication in R&D assume that R&D is undertaken to create information. But, as Pavitt (1987) points out, if the R&D expenditure is undertaken to assimilate existing technological knowledge, then duplication may not be wasteful, but may instead be essential for diffusion.

Information as a Stock which Appreciates with Use

Information has some of the characteristics of a capital good, in the sense that it is more like a stock that yields services rather than a flow which disappears after consumption. Yet unlike other capital goods, it does not depreciate with use; indeed, as Dasgupta and Stoneman (1987) amongst others have noted, information is a stock that can *appreciate* with use. By this, we mean that the quality of the information – i.e. its accuracy and conciseness – appreciates with use. On the other hand, the competitive value of information does not usually increase with use because competitive edge depends on a degree of exclusivity: if potentially valuable information is shared too widely, its competitive value declines.

The quality of information improves with more frequent use by the 'owner' of that information, and in addition the quality of information is improved when it is passed around a network of users who can hone and refine it. A clear example of this can be seen with *Wikipedia*, the open source encyclopedia, where users are encouraged to update and improve information. Because of this, information gathering, exchange, and use are subject to special and rich network effects: these are positive mutual benefits between collectors and users of information.

The fact that a stock of information appreciates as it is used by other firms can imply a mutual positive externality. Market failure may manifest itself in the form of information networks that are *too small* because agents deciding whether or not to join are not rewarded for the externalities they may generate by being part of that network.

Information Asymmetries

It is often argued that information has one very special property that is not shared by most other goods: namely that the value of information to a particular user often cannot be assessed without having temporary access to the information (on a trial basis, as it were). But when the buyer has had access to the information (even on a temporary basis), there is no need to buy. And after access, there is no incentive for the buyer to reveal his/her true willingness-to-pay for the information. The argument implies, therefore, that information cannot really be described without 'giving it away'. In such cases, there is an informational asymmetry between buyer and seller, and an element of moral hazard. In such circumstances, markets will fail.

This is undoubtedly true of many kinds of information. Thus the potential value of a document giving (for example) the detailed formula of a revolutionary new pharmaceutical product cannot really be assessed from the statement: 'this document gives the precise formula of a revolutionary new pharmaceutical product'. But this phenomenon is not unique to information. It can indeed be difficult to assess the value of all sorts of goods before purchase, even if trial periods are offered – and this is particularly true in an environment of very rapid change. The marketing and sale of all sorts of commodities requires the use of language to convey relevant information about the product (Bacharach, 1990), without 'giving it away' free, and while language may be fairly well adapted to describe a car, a computer, or a cake, there is still an element of incompleteness about the description, and a degree of information asymmetry.

With information asymmetries, we find that the seller knows the potential value of a piece of information, while the buyer is uncertain about it. In such cases, a market for information may not work efficiently. The best known examples of this are Gresham's Law and the *lemon effect* described by Akerlof (1970). The existence of bad information (of little value) reduces the market clearing price, which may in turn lead the supplier of good information to withdraw his/her supply so further reducing the average value of available information. Sometimes the existence of reputation effects will be sufficient to ensure that 'good' information (from a trusted supplier) can trade at the necessary premium. And this is why, as Metcalfe (1986) points out, seller reputation and buyer goodwill are so important in information exchange.

GENERIC APPROACHES TO POLICY

Market failure may be a necessary condition to justify government having an innovation policy, but it is not a sufficient reason. It is necessary that government can find policy devices which might actually improve upon the market outcome and in a cost-effective way. This is not unproblematic. First, there is the whole matter of appropriate policies in a second best world. We may know what the optimum compensatory policy will be when there is only one market imperfection (e.g. one source of externalities), but in a world where there are many inter-dependent market imperfections, then we can no longer say that this is the optimal policy.

In a world of very poor information, with little or no prior knowledge about the full extent of market failure, some would argue that the best policy is no intervention, even if one market imperfection can be clearly identified.

As one example of this difficulty, Stoneman (1987) examines the effects of subsidies and information provision schemes on diffusion. He shows that depending on the expectations of users and the relevant market structure, the market may generate a diffusion path that is *too slow* or *too fast* from a social welfare point of view. So in these circumstances, it is not clear whether information provision and/or adoption subsidies are appropriate.

It is also important to note at the outset that when a programme of government intervention is being evaluated, the policies are unlikely to appear profitable in an accounting sense. The reason for this is obvious enough. The policies are trying to make certain activities happen that are *not privately* profitable but are nevertheless *socially* profitable.

Nevertheless, all these possible problems in designing effective innovation policies have not deterred most governments, and we find in practice that there are many policies designed to increase innovation in the economy. Dasgupta (1987, 1988) distinguishes three broad groups of government policy to correct for market failure in innovation:

1) subsidies to reflect the positive externalities arising from R&D activities and other innovation activities
2) institutions to create and enforce property rights to ensure there are no free spillovers to third parties
3) government expenditure or procurement to promote activities that the market fails to support adequately.

The subsidy approach is sometimes called the *Pigou approach* since it follows in the tradition of policies advocated by the early twentieth-century economist, Pigou. The system of property rights is sometimes called the *Lindahl approach*, after the Swedish economist Lindahl. And when government gets directly involved in R&D and other innovation activities, that is sometimes called the *Samuelson approach* after the Nobel-Prize winning economist, Paul Samuelson.

SUBSIDIES

These can take two forms: subsidies on provision or subsidies on use. Subsidies raised from general taxation are paid to compensate those who create positive externalities through their R&D for those externalities. Or, in the case of subsidising diffusion, the user is subsidised to adopt a new technology to compensate him/her for the positive externality (s)he creates in adopting the technology.

Subsidies can be general or specific. General subsidies, such as tax breaks on R&D expenditure are administratively fairly straightforward, but their effects can be to promote a range of additional activities – not all of which the government is keen to encourage. Specific subsidies, by contrast, might be given for particularly promising areas of R&D or to promote adoption of a particularly promising new technology. These are administratively more complex and costly, but their effects will be more clearly focused on a specific range of activities which the government is *indeed* keen to encourage.

The main question to be asked of such schemes is whether they promote *additionality*. That is, to what extent is *additional* R&D activity (or technology adoption) generated by such schemes? Or, alternatively, do the subsidies simply turn into an increased surplus for those who are already happily undertaking an R&D project or are already using a new technology? In short, how efficient are the subsidies in terms of extra social benefits generated per pound spent in subsidy? Some sceptics in recent years have argued that such schemes have low additionality and as a result these schemes are less popular now than they were some 25 years ago.

Practical Policies

In the UK, the government has experimented with a variety of policies of this sort:

- R&D tax credits are the biggest single funding mechanism for business R&D provided by the British government.[1] R&D tax credits help R&D-active companies to reduce their tax bill or, in the case of small or medium sized companies (SMEs) who are not making profits, and who are therefore paying no tax, by providing a cash sum.
- Knowledge Transfer Networks offer financial support to organisations that have the capability to establish or enhance networks. This can be seen as a subsidy paid in return for creating positive spillovers.
- Direct grants for R&D are available to startup companies and SMEs to carry out research and development work on technologically innovative products and processes.
- Programmes to support specific technologies (electronics, biotechnology).
- Technology adoption or diffusion subsidies.
- Technology training subsidies and consultancy subsidies.

PROPERTY RIGHTS

Such schemes require the externalities from R&D activity and information gathering to be 'commoditised' by use of intellectual property rights. We have already discussed the different approaches to protecting intellectual property in Chapter 7. When these externalities are commoditised in this way, the inventor is in a position to act as a licenser of the spillovers from his/her activities, and thus to internalise some of the externalities that would otherwise be enjoyed free by others.

The argument for doing this is clear enough. But there is a downside. The cost of a patent system, besides the administrative cost, is that it gives monopoly rights to inventors; as with any form of monopoly, that is inefficient. Moreover, such a scheme acts against technology diffusion by putting a price on something that was originally a free spillover.

Once again we see the inherent dilemma in technology policy. Incomplete appropriation leads to insufficient provision under the market solution, and the establishment of property rights can help to achieve a higher level of provision. But the establishment of property rights over something that under natural conditions would spill over at zero marginal cost will itself introduce an element of inefficiency.

Practical Policies

Institutions for protecting intellectual property have existed for some time (see Chapter 8). The UK government's main policy in this area has been to commission a wide-ranging review of the intellectual property framework, called the Gowers Review (2006). This review recognised that intellectual property is a critical component of the UK's strategy for success in the global economy, but needed some reform. The key challenge for the IP framework is to create incentives for innovation, without unduly limiting access for consumers and follow-on innovators. The Review recognised three priorities:

- tackling IP crime and ensuring that IP rights are well enforced
- reducing the costs and complexity of the system
- reforming copyright law to allow individuals and institutions to use content in ways consistent with the digital age.

Following the Gowers Review, and in recognition of a changing balance between different instruments for protecting IP, the UK *Patent Office* was given a new name: the UK *Intellectual Property Office*.[2]

GOVERNMENT'S OWN R&D ACTIVITIES

In severe cases of market failure, the government can in principle assure that socially valuable (but privately unprofitable) activities take place by direct government expenditure on R&D, or other expenditures on the science and technology infrastructure.

The first obvious difficulty with direct public-sector involvement in R&D is that public agencies lack the commercial information and market incentives which firms in the industry would possess. For this reason, it is generally argued that public involvement in R&D is more appropriate with basic research than with near-market R&D. And moreover, it is often argued that positive externalities and the appropriability problem are more important in basic research than in near-market research.

The second possible difficulty is that government funded research could crowd out privately funded research. This might happen as a consequence of competing for scarce scientific or technological talent. Or it might simply displace the incentive for privately funded research, by freely offering the results of such research to companies (that would be willing to pay for it on their own). Another way of looking at this is to recognise the possibility that government funding of R&D perpetuates a dependent corporate culture in which research is, 'something the government pays for'.

Practical Policies

In the UK, the government has a wide variety of policies of this third sort. Indeed, as the number of subsidy schemes has declined because of concerns about additionality (see above) this number of schemes of this third sort has grown. All of these policies can be seen as an attempt to add to and strengthen the group of scientific and technology institutions that make up the *national system of innovation* in the UK. We can group these policies into three sorts, as shown in this final list:

- Infrastructure
  - Government research laboratories
  - Standards institutions and measurement laboratories
  - Promoting clusters
  - Technology transfer institutions

- Education and training
  - Direct sponsorship of university research
  - Sponsorship for collaboration between universities and industry
  - Engineering and technology programmes
  - The 'Micros in Schools' programme

- Vision and foresight
  - Foresight programmes
  - The 'Technology Strategy Board'

INNOVATION POLICY IN THE FUTURE

I finish this chapter, and indeed the book, with a conjecture. In the future, government policy towards innovation will be different from policy towards innovation in the past and present.

Why do I say this? There are two main reasons. First, much policy is still governed by a relatively simplistic model of how innovation happens and how innovation helps to create wealth. A common argument is that invention and creativity don't really count until they turn into innovation, and innovation doesn't really count until it impacts on company productivity and/or profitability. Chapter 19 would suggest that this perspective is far too narrow. When we take account of the multiple channels through which creativity, invention and innovation can create wealth, then a more subtle approach to policy is required.

Second, most past policy seems to be governed by the assumption that more innovation is always good. As one policy maker said to me, the message has been, in essence: 'go forth and innovate!' It is only a slight over-statement to say that the main object of policy is to increase the amount of innovation – more or less uncritically. But again, we have seen in Chapters 19 and 21 that more innovation is not always a good thing.

What will this new approach to policy look like? The simple answer is that we don't know yet. But it will be more subtle than the approaches described above.

In conclusion, let us remind ourselves of the remark by Ernst Schumacher (1974, p. 26) noted in Chapter 2: 'man is far too clever to be able to survive without wisdom'. We have the power to come up with all kinds of clever innovations and many of them may enhance the competitiveness and performance of the companies that implement them. But the full implications of these innovations for sustainability and welfare may be a great deal more complicated than the immediately obvious effects, and some apparently benign innovations can have unexpected and damaging side-effects. The policy objective must move away from how to achieve *more and more* innovation, of whatever kind, to how to achieve more of the *right sort* of innovation.

NOTES

[1] http://www.berr.gov.uk/dius/innovation/randd/
[2] http://www.ipo.gov.uk/

Appendix: supplementary reading

As indicated at the start, the emphasis in this introductory text has been on breadth rather than depth. The students and tutors who wish to explore these topics in more depth may wish to consult the following readings.

General Texts on Economics of Innovation

Antonelli, C. (2002), *The Economics of Innovation, New Technologies and Structural Change*, London: Routledge

Arora, A., A. Fosfuri and A. Gambardella (2001), *Markets for Technology: The Economics of Innovation and Corporate Strategy*, Cambridge, MA: MIT Press

Dosi, G. (1988), 'Sources, Procedures, and Microeconomic Effects of Innovation', *Journal of Economic Literature*, **26**, 1120-1171

Fagerberg, J., D.C. Mowery, and R.R. Nelson (2006), *The Oxford Handbook of Innovation*, Oxford: Oxford University Press

Foray, D. (2004), *The Economics of Knowledge*, Cambridge, MA: MIT Press

Freeman, C. and L. Soete (1997), *The Economics of Industrial Innovation*, 3rd edition, London: Continuum

Scotchmer, S. (2004), *Innovation and Incentives*, Cambridge, US: MIT Press

Mansell, R. and W.E. Steinmueller (2000), *Mobilizing the Information Society: Strategies for Growth and Opportunity*, Oxford: Oxford University Press

Stoneman, P. (1983), *The Economic Analysis of Technological Change*, Oxford: Oxford University Press

Stoneman, P. (ed.) (1995), *Handbook of the Economics of Innovation and Technological Change*, Oxford: Blackwell Publishers

History of Economic Thought

Blaug, M. (1997), *Economic Theory in Retrospect*, Cambridge: Cambridge University Press

Schumpeter, J. (1994), *History of Economic Analysis*, with an introduction by M. Perlman, London: Routledge

Process Innovation

Fagerberg, J., D.C. Mowery, and R.R. Nelson (2006), *The Oxford Handbook of Innovation*, Oxford: Oxford University Press

Stoneman, P. (1983), *The Economic Analysis of Technological Change*, Oxford: Oxford University Press

Product Innovation

Swann, G.M.P. (1986), *Quality Innovation: An Economic Analysis of Rapid Improvements in Microelectronic Components*, London: Frances Pinter

Trajtenberg, M. (1990), *Economic Analysis of Product Innovation: The Case of CT Scanners*, Cambridge, MA: Harvard University Press

Innovative Pricing

Jonason, A. (2001), *Innovative Pricing*, PhD Dissertation, Royal Institute of Technology (KTH), Stockholm, available at http://www.diva-portal.org/kth/

Phlips, L. (1983), *The Economics of Price Discrimination*, Cambridge: Cambridge University Press

Network Effects and Standards

David, P.A. (1985), 'Clio and the Economics of QWERTY', *American Economic Review*, **75**, 332-336

Grindley, P.C. (1995), *Standards Strategy and Policy: Cases and Stories*, Oxford: Oxford University Press

Rohlfs, J.H. (2001), *Bandwagon Effects in High Technology Industries*, Cambridge, MA: MIT Press

Shy, O. (2001), *The Economics of Network Industries*, Cambridge: Cambridge University Press

Intellectual Property

Chesborough, H. (2003), *Open Innovation: The New Imperative for Creating and Profiting from Technology*, Boston, MA: Harvard Business School Press

Granstrand, O. (1999), *The Economics and Management of Intellectual Property*, Cheltenham, UK and Northampton, MA, USA: Edward Elgar

Jaffe, A.B. and J. Lerner (2004), *Innovation and Its Discontents: How Our Broken Patent System is Endangering Innovation and Progress, and What to Do About It*, Princeton, NJ: Princeton University Press

Scotchmer, S. (2004), *Innovation and Incentives*, Cambridge, MA: MIT Press

Theories of Creativity

Csikszentmihalyi, M. (1996), *Creativity: Flow and the Psychology of Discovery and Invention,* New York: Harper Perennial

Rickards, T. (1999), *Creativity and the Management of Change*, Oxford: Blackwell Publishers

Runco, M.A. and S.R. Pritzker (eds) (1999), *Encyclopaedia of Creativity*, Volumes 1-2, San Diego, CA: Academic Press

The Entrepreneur

Earl, P.E. and T. Wakeley (2005), *Business Economics: A Contemporary Approach*, Maidenhead, UK: McGraw-Hill

Earl, P.E. (2003), 'The Entrepreneur as a Constructor of Connections', in R. Koppl (ed.), *Advances in Austrian Economics*, **6**, 117-134, Amsterdam: JAI/Elsevier

Link, A.N. and D.S. Siegel (2007), *Innovation, Entrepreneurship, and Technological Change*, Oxford: Oxford University Press

Schumpeter, J.A. (1954), *Capitalism, Socialism and Democracy*, 4th Edition, London: Unwin University Books

Organisation for Innovation

Nelson, R.R. and S.G. Winter (1982), *An Evolutionary Theory of Economic Change*, Cambridge, MA: The Belknap Press of Harvard University

Stamm, B. von (2003), *Managing Innovation, Design and Creativity*, Chichester, UK: John Wiley and Sons

Tidd, J., J. Bessant and K. Pavitt (2001), *Managing Innovation: Integrating Technological, Market and Organisational Change*, 2nd edition, Chichester, UK: John Wiley and Sons

Technology Vision

Dosi, G. (1982), 'Technological Paradigms and Technological Trajectories: A Suggested Interpretation of the Determinants and Directions of Technical Change', *Research Policy*, **11**, 147-162

Hamel, G. and C.K. Prahalad (1994), *Competing for the Future*, Boston, MA: Harvard University Press

Swann, G.M.P. and J. Gill (1993), *Corporate Vision and Rapid Technological Change*, London: Routledge

Clusters and Networks

Martin, R. and P. Sunley (2003), 'Deconstructing Clusters: Chaotic Concept or Policy Panacea?', *Journal of Economic Geography*, **3**, 5-35

Porter, M. (1990), *The Competitive Advantage of Nations*, London: Macmillan

Swann, G.M.P., M. Prevezer and D. Stout (eds) (1998), *The Dynamics of Industrial Clustering*, Oxford: Oxford University Press

Clark, G.L., M.S. Gertler and M.P. Feldman (eds) (2000), *The Oxford Handbook of Economic Geography*, Oxford: Oxford University Press

Division of Labour

Smith, A. (1776/1904), *An Inquiry into the Nature and Causes of the Wealth of Nations*, Volumes I and II, London: Methuen

Babbage, C. (1835), *On the Economy of Machinery and Manufactures*, 4th edition, London: Charles Knight

Durkheim, E. (1893/1984), *The Division of Labour in Society*, translated by W.D. Halls, New York: Free Press

Innovation and the Consumer

Becker, G.S. (1996), *Accounting for Tastes*, Cambridge, MA: Harvard University Press

Bianchi, M. (1998), *The Active Consumer: Novelty and Surprise in Consumer Choice*, London: Routledge

Hippel, E. von (2005), *Democratizing Innovation*, Cambridge, MA: MIT Press

McMeekin, A., K. Green, M. Tomlinson and V. Walsh (eds) (2002), *Innovation by Demand: An Interdisciplinary Approach to the Study of Demand and its Role in Innovation*, Manchester: Manchester University Press

The Diffusion of Innovations

Baptista, R. (1999), 'The Diffusion of Process Innovations: A Selective Review', *International Journal of the Economics of Business*, **6**, 107-129

Geroski, P.A. (2000), 'Models of Technology Diffusion', *Research Policy*, **29**, 603-625

Rogers, E. (1995), *Diffusion of Innovations*, 4th edition, New York: Free Press

Stoneman, P. (2002), *The Economics of Technological Diffusion*, Oxford: Blackwell

Innovation and Trade

Cairncross, F. (1997), *The Death of Distance*, Boston, MA: Harvard Business School Press

Krugman, P. (1991), *Geography and Trade*, Cambridge, MA: MIT Press

Sheshinski, E., R.J. Strom and W.J. Baumol (eds) (2007), *Entrepreneurship, Innovation, and the Growth Mechanism of the Free-Enterprise Economies*, Princeton, NJ: Princeton University Press

Innovation and Market Structure

Kamien, M.I. and N.L. Schwartz (1982), *Market Structure and Innovation*, Cambridge: Cambridge University Press

Cohen, W.M. and R.C. Levin (1989), 'Empirical Studies of Innovation and Market Structure', in R. Schmalensee and R. Willig (eds), *Handbook of Industrial Organization*, Volume 2, Amsterdam: Elsevier

Mazzucato, M. (2000), *Firm Size, Innovation and Market Structure: The Evolution of Market Concentration and Instability*, Cheltenham, UK and Northampton, MA, USA: Edward Elgar

Sutton, J. (1998), *Technology and Market Structure*, Cambridge, MA: MIT Press

Innovation and Wealth Creation

Baumol, W.J. (2002), *The Free-Market Innovation Machine: Analyzing the Growth Miracle of Capitalism*, Princeton, NJ: Princeton University Press

Landes, D. (2003), *The Unbound Prometheus: Technological Change and Industrial Development in Western Europe from 1750 to the Present*, 2nd edition, Cambridge: Cambridge University Press

Metcalfe, J.S. (1998), *Evolutionary Economics and Creative Destruction*, London: Routledge

Rosenberg, N., R. Landau and D.C. Mowery (1992), *Technology and the Wealth of Nations*, Stanford, CA: Stanford University Press

Tunzelmann, G.N. von (1995), *Technology and Industrial Progress: The Foundations of Economic Growth*, Aldershot, UK and Brookfield, USA: Edward Elgar

Innovation and Competitiveness

Geroski, P.A. and C. Markides (2004), *Fast Second: How Smart Companies Bypass Radical Innovation to Enter and Dominate New Markets*, San Francisco, CA: Jossey-Bass
Mytelka, L.K. (1999), *Competition, Innovation and Competitiveness in Developing Countries*, Paris: OECD
Utterback, J.M. (1994), *Mastering the Dynamics of Innovation*, Boston, MA: Harvard Business School Press

Innovation and Sustainability

Elzen, B., F.W. Geels and K. Green (2004), *System Innovation and the Transition to Sustainability: Theory, Evidence and Policy*, Cheltenham, UK and Northampton, MA, USA: Edward Elgar
OECD (2000), *Innovation and the Environment*, OECD Working Group on Innovation and Technology Policy, Paris: OECD
Schumacher, E. (1974), *Small is Beautiful*, London: Abacus/Sphere Books
Stern Review on the Economics of Climate Change (2006), available at http://www.hm-treasury.gov.uk/sternreview_index.htm

Government Policy for Innovation

Archibugi, D., J. Howells and J. Michie (1999), *Innovation Policy in a Global Economy*, Cambridge: Cambridge University Press
DIUS (2008), *Innovation Nation*, Science and Innovation White Paper, 13 March 2008, available at http://www.dius.gov.uk/
Metcalfe, J.S. (1995), 'The Economic Foundations of Technology Policy: Equilibrium and Evolutionary Perspectives', in P. Stoneman (ed.), *Handbook of the Economics of Innovation and Technological Change*, Oxford: Blackwell Publishers
Nelson, R. (ed.) (1993), *National Innovation Systems. A Comparative Analysis*, Oxford: Oxford University Press
Stoneman, P. (1987), *The Economic Analysis of Technology Policy*, Oxford: Oxford University Press

References

Abra, J. and G. Abra (1999), 'Collaboration and Competition', in M.A. Runco and S.R. Pritzker (eds), *Encyclopedia of Creativity: Volume 1*, San Diego, CA: Academic Press, pp. 283-94

Akerlof, G. (1970), 'The Market for Lemons', *Quarterly Journal of Economics*, **84**, 488-500

Altschuler, A., M. Anderson, D.T. Jones, D. Roos and J. Womack (1985), *The Future of the Automobile*, Cambridge, MA: MIT Press

Amabile, T.M. (1996), *Creativity in Context: Update to the Social Psychology of Creativity*, Boulder, CO: Westview Press

Aoki, M. (1986), 'Horizontal versus Vertical Information Structure of the Firm', *American Economic Review*, **76**, 971-983

Arrow, K.J. (1962), 'Economic Welfare and the Allocation of Resources for Inventions', in R.R. Nelson (ed.), *The Rate and Direction of Inventive Activity*, Princeton, NJ: Princeton University Press

Audretsch, D. (1992), 'The Technological Regime and Market Evolution: The New Learning', *Economics of Innovation and New Technology*, **2**, 27-35

Babbage, C. (1835), *On the Economy of Machinery and Manufactures*, 4th edition, London: Charles Knight

Bacharach, M. (1990), 'Commodities, Language, and Desire', *The Journal of Philosophy*, **87**, 346-368

Barley, S.R. (1986), 'Technology as an Occasion for Structuring: Evidence from Observation of CT Scanners and the Social Order of Radiology Departments', *Administrative Science Quarterly*, **31**, 78-108

Barron, F. (1963), 'The Needs for Order and for Disorder as Motives in Creative Activity', in C.W. Taylor and F. Barron (eds), *Scientific Creativity: Its Recognition and Development*, New York: Wiley

Becker, G.S. (1996), *Accounting for Tastes*, Cambridge, MA: Harvard University Press

Bernsen, J. (1987), 'Design in Action', in *Design Management in Practice*, Copenhagen: Danish Design Council

Birke, D. and G.M.P. Swann (2005), 'Social Networks and Choice of Mobile Phone Operator', Occasional Paper 14, Nottingham University Business School, http://www.nottingham.ac.uk/~lizecon/RePEc/pdf/networks.pdf

Blair, J.M. (1948), 'Technology and Size', *American Economic Review*, **38**, 121-152

Blair, J.M. (1972), *Economic Concentration: Structure, Behaviour and Public Policy*, New York: Harcourt, Brace, Jovanovich

Bourdieu, P. (1984), *Distinction: A Social Critique of the Judgment of Taste*, London: Routledge & Kegan Paul

Brewer, A. (1998), 'Invention', in O.F. Hamouda, C. Lee and D. Mair (eds), *The Economics of John Rae*, London: Routledge

Burns, T. and G. Stalker (1961), *The Management of Innovation*, London: Tavistock

Cairncross, F. (1997), *The Death of Distance*, Boston, MA: Harvard Business School Press

Campbell, R. (1747/1969), *The London Tradesman*, Facsimile reprint of 1st edition, Newton Abbot, UK: David and Charles

Chandler, A.D. Jr (1962), *Strategy and Structure*, Cambridge, MA: MIT Press

Chandler, A.D. Jr (1977), *The Visible Hand: The Managerial Revolution in American Business*, Cambridge, MA: Belknap Press

Child, J. (1984), *Organisation: A Guide to Problems and Practice*, 2nd edition, London: Harper & Row

Cipolla, C. (1967), *Clocks and Culture: 1300-1700*, London: William Collins and Sons

Clay, K., R. Krishnan and E. Wolff (2001), 'Prices and Price Dispersion on the Web: Evidence from the Online Book Industry', NBER Working Paper 8271, Cambridge, MA: NBER

Cohen, W.M. and D.A. Levinthal (1989), 'Innovation and Learning: The Two Faces of R&D', *Economic Journal*, **99**, 569-596

Cohen, W.M. and D.A. Levinthal (1990), 'Absorptive Capacity: A New Perspective on Learning and Innovation', *Administrative Science Quarterly*, **35**, 128-152

Comanor, W.S. (1964), 'Research and Competitive Product Differentiation in the Pharmaceutical Industry in the United States', *Economica*, **31**, 372-384

Cook, E.T. and A. Wedderburn (eds) (1903-1912/1996), *The Works of John Ruskin*, CD-ROM Version, Cambridge: Cambridge University Press

Cowan, R., W. Cowan and G.M.P. Swann (2004), 'Waves in Consumption with Interdependence Between Consumers', *Canadian Journal of Economics*, **37**, 149-177

Daft, R.L. (1982), 'Bureaucratic versus Non-Bureaucratic Structure and the Process of Innovation and Change', in S.B. Bacharach (ed.), *Research in the Sociology of Organisations*, **1**, 129-166, Greenwich, CT: JAI Press

Dalle, J.-M. and P.A. David (2007), 'It Takes All Kinds: A Simulation Modelling Perspective on Motivation and Coordination in Libre Software

Development Projects', SIEPR Discussion Paper No. 07-24, Stanford Institute for Economic Policy Research

Dasgupta, P. (1987), 'The Economic Theory of Technology Policy', in P. Dasgupta and P. Stoneman (eds), *Economic Policy and Technological Performance*, Cambridge: Cambridge University Press

Dasgupta, P. (1988), 'The Welfare Economics of Knowledge Production', *Oxford Review of Economic Policy*, **4**, 1-12

Dasgupta, P. and E. Maskin (1987), 'The Simple Economics of Research Portfolios', *Economic Journal*, **97**, 581-595

Dasgupta, P. and J.E. Stiglitz (1980a), 'Uncertainty, Industrial Structure and the Speed of R&D', *Bell Journal of Economics*, **11**, 1-28

Dasgupta, P. and J.E. Stiglitz (1980b), 'Industrial Structure and the Nature of Innovative Activity', *Economic Journal*, **90**, 266-293

Dasgupta, P. and P. Stoneman (1987), 'Introduction', in P. Dasgupta and P. Stoneman (eds), *Economic Policy and Technological Performance*, Cambridge: Cambridge University Press

David, P.A. (1985), 'Clio and the Economics of QWERTY', *American Economic Review*, **75**, 332-336

David, P.A. (1997), 'Path Dependence and the Quest for Historical Economics: One More Chorus of the Ballad of QWERTY', Discussion Papers in Economic and Social History, Number 20, University of Oxford

Design Council (1995), *Definitions of Design*, London: Design Council

Dogan, M. (1994), 'Fragmentation of the Social Sciences and Re-Combination of Specialties', *International Social Science Journal*, **139**, 27-42

Dogan, M. and R. Pahre (1990), *Creative Marginality: Innovation at the Intersection of Social Sciences*, Boulder, CO: Westview Press

Dosi, G. (1984), *Technical Change and Industrial Transformation*, London: Macmillan

Douglas, M. (1983), 'Identity: Personal and Socio-Cultural', *Uppsala Studies in Cultural Anthropology*, **5**, 35-46

Downie, J. (1958), *The Competitive Process*, London: Duckworth

Draca, M., R. Sadun and J. van Reenan (2007), 'Productivity and ICTs: A Review of the Evidence', in R. Mansell, C. Avgerou, D. Quah and R. Silverstone (eds), *The Oxford Handbook of Information and Communication Technologies*, Oxford: Oxford University Press, pp. 100-147

Duesenberry, J.S. (1960), 'Comment on an Economic Analysis of Fertility', in Universities – National Bureau Committee for Economic Research (eds), *Demographic and Economic Change in Developed Countries: A Conference*, Princeton, NJ: Princeton University Press, for the National Bureau of Economic Research, pp. 231-234

Durkheim, E. (1893/1984), *The Division of Labour in Society*, translated by W.D. Halls, New York: Free Press

Dutton, J. and A. Thomas (1985), 'Relating Technological Change and Learning by Doing', in R.D. Rosenbloom (ed.) *Research on Technological Innovation, Management and Policy*, **2**, 187-224, Greenwich, CT: JAI Press

Earl, P.E. (2003), 'The Entrepreneur as a Constructor of Connections', in R. Koppl (ed.), *Advances in Austrian Economics*, **6**, 117-134, Amsterdam: JAI/Elsevier

Earl P.E. and T. Wakeley (2005), *Business Economics: A Contemporary Approach*, Maidenhead, UK: McGraw-Hill

Economist (2000), 'Internet Economics: A Thinker's Guide', *The Economist*, 30 March

Ekvall, G. (1987), 'The Climate Metaphor in Organizational Theory', in B.M. Bass and P.J.D. Drenth (eds), *Advances in Organizational Psychology*, Beverly Hills, CA: Sage, pp. 177-190

Ekvall, G. (1996), 'Organisational Climate for Creativity and Innovation', *European Journal of Work and Organisational Psychology*, **5**, 105-123

Farrell, J. and G. Saloner (1985), 'Standardization, Compatibility and Innovation', *RAND Journal of Economics*, **16**, 70-83

Fisher, F.M., J.J. McGowan and J.E. Greenwood (1983), *Folded, Spindled and Mutilated: Economic Analysis and U.S. vs. IBM*, Cambridge, MA: MIT Press

Freeman, C. (1982), *The Economics of Industrial Innovation*, 2nd edition, London: Frances Pinter

Freeman, C. (2007), 'The ICT Paradigm', in R. Mansell, C. Avgerou, D. Quah and R. Silverstone (eds), *The Oxford Handbook of Information and Communication Technologies*, Oxford: Oxford University Press

Freeman, C., C.J.E. Harlow and J.K. Fuller (1965), 'Research and Development in Electronic Capital Goods', *National Institute Economic Review*, **34**, 40-91

Freeman, C. and L. Soete (1997), *The Economics of Industrial Innovation*, 3rd edition, London: Continuum

Gabel, H.L. (1991), *Competitive Strategies for Product Standards: The Strategic Use of Compatibility Standards for Competitive Advantage*, London: McGraw-Hill

Galbraith, J.K. (1958), *The Affluent Society* Boston, MA: Houghton Mifflin

Gardner, H. and C. Wolf (1988), 'The Fruits of Asynchrony: A Psychological Examination of Creativity', *Adolescent Psychiatry*, **15**, 96-120

George, H. (1879/1931), *Progress and Poverty*, London: The Henry George Foundation of Great Britain

Geroski, P.A. (1990), 'Innovation, Technological Opportunity and Market Structure', *Oxford Economics Papers*, **42**, 586-602

Geroski, P.A. and C. Markides (2004), *Fast Second: How Smart Companies Bypass Radical Innovation to Enter and Dominate New Markets*, San Francisco, CA: Jossey-Bass

Geroski, P.A. and R. Pomroy (1990), 'Innovation and Evolution of Market Structure', *Journal of Industrial Economics*, **38**, 299-314

GLA Economics (2002), *Creativity: London's Core Business*, available at http://www.london.gov.uk/gla/publications/economy.jsp

Gort, M. and Klepper, S. (1982), 'Time Paths in the Diffusion of Product Innovations', *Economic Journal*, **92**, 630-653

Gowers Review of Intellectual Property (2006), available at http://www.hm-treasury.gov.uk/gowers_review_index.htm

Griliches, Z. (1957), 'Hybrid Corn: An Exploration in the Economics of Technological Change', *Econometrica*, **25**, 501-522

Grindley, P.C. (1995), *Standards Strategy and Policy: Cases and Stories*, Oxford: Oxford University Press

Grubel, H.G. and P.J. Lloyd (1975), *Intra-industry Trade: The Theory and Measurement of International Trade in Differentiated Products*, London: Macmillan

Guardian (2008), 'Sony's Blu-ray wins HD DVD Battle', *The Guardian*, 19 February, available at: http://www.guardian.co.uk

Hall, B. (2005), 'Exploring the Patent Explosion', *Journal of Technology Transfer*, **30**, 35-48

Hannan, M.T. and J. Freeman (1977), 'The Population Ecology of Organisations', *American Journal of Sociology*, **83**, 929-964

Hannan, M.T. and J. Freeman (1984), 'Structural Inertia and Organisational Change', *American Sociological Review*, **49**, 149-164

Henderson, R.M. and K.B. Clark (1990), 'Architectural Innovation: The Reconfiguration of Existing Product Technologies and the Failure of Established Firms', *Administrative Science Quarterly*, **35**, 9-30

Hippel, E. von (1988), *The Sources of Innovation*, Oxford: Oxford University Press

Hippel, E. von (2005), *Democratizing Innovation*, Cambridge, MA: MIT Press

Hirsch, S. (1965), 'The United States Electronics Industry in International Trade', *National Institute Economic Review*, **34**, 92-97

Horowitz, I. (1962), 'Firm Size and Research Activity', *Southern Economic Journal*, **28**, 298-301

Hotelling, H. (1929), 'Stability in Competition', *Economic Journal*, **39**, 41-57

Hudson, L. (1966), *Contrary Imaginations*, New York: Schocken Books

Hutton, W. (2000), 'No IT, No Comment', *The Observer*, 19 November

Janis, I. (1972), *Victims of Groupthink*, Boston, MA: Houghton Mifflin

Janis, I. (1982), *Groupthink: Psychological Studies of Policy Decisions and Fiascos*, 2nd edition, Boston, MA: Houghton Mifflin

Jevons, W.S. (1878), *Political Economy*, London: Macmillan

Karshenas, M. and P. Stoneman, P. (1993), 'Rank, Stock, Order, and Epidemic Effects in the Diffusion of New Process Technologies – An Empirical Model', *Rand Journal of Economics*, **24**, 503-528

Katz, B. and Phillips, A. (1982), 'Innovation, Technological Change and the Emergence of the Computer Industry', in H. Giersch (ed.), *Emerging Technology*, Tubingen: J.C.B. Mohr

Katz, M.L. and C. Shapiro (1985), 'Network Externalities, Competition and Compatibility', *American Economic Review*, **75**, 424-440

Kay, N.M. (1997), *Patterns in Corporate Evolution*, Oxford: Oxford University Press

Kirzner, I. (1979), *Perception, Opportunity and Profit: Studies in the Theory of Entrepreneurship*, Chicago, IL: University of Chicago Press

Klemperer, P. (1987), 'Markets with Consumer Switching Costs', *Quarterly Journal of Economics*, **102**, 375-394

Klepper, S. (1996), 'Entry, Exit, Growth, and Innovation over the Product Life Cycle', *American Economic Review*, **86**, 562-583

Koestler, A. (1964), *The Act of Creation*, London: Hutchinson

Kotler, P., L. Fahey and S. Jatusripitak (1986), *The New Competition: Meeting the Marketing Challenge from the Far East*, Englewood Cliffs, NJ: Prentice Hall International

Krugman, P. (1991), *Geography and Trade*, Cambridge, MA: MIT Press

Laffont, J.-J., P. Rey and J. Tirole (1998), 'Network Competition: II. Price Discrimination', *RAND Journal of Economics*, **29**, 38-56

Lancaster, K.J. (1971), *Consumer Demand: A New Approach*, New York: Columbia University Press

Landes, D. (1983), *Revolution in Time*, Cambridge, MA: Belknap Press of Harvard University Press

Langlois, R.N. (1992), 'External Economies and Economic Progress: The Case of the Microcomputer Industry', *Business History Review*, **66**, 1-50

Leontief, W. (1953), 'Domestic Production and Foreign Trade; The American Capital Position Re-examined', *Proceedings of the American Philosophical Society*, **97**, 332-349

Leontief, W. (1956), 'Factor Proportions and the Structure of American Trade: Further Theoretical and Empirical Analysis', *The Review of Economics and Statistics*, **38**, 386-407

Lerner, J. and J. Tirole (2002), 'Some Simple Economics of Open Source, *Journal of Industrial Economics*, **50**, 197-234

McCulloch, J.R. (1864/1965), *The Principles of Political Economy*, Reprints of Economic Classics, New York: Augustus Kelley

McKelvey, B. and Aldrich, H. (1983), 'Populations, Natural Selection, and Applied Organisational Science', *Administrative Science Quarterly*, **28**, 101-128

Malerba, F. (1985), *The Semiconductor Business: The Economics of Rapid Growth and Decline*, London: Frances Pinter

Mansfield, E. (1961), 'Technical Change and the Rate of Imitation', *Econometrica*, **29**, 741-766

Mansfield, E. (1962), 'Entry, Gibrat's Law, Innovation and the Growth of Firms', *American Economic Review*, **52**, 1023-1051

Mansfield, E. (1968a), *Industrial Research and Technical Innovation*, New York: Norton

Mansfield, E. (1968b), *The Economics of Technical Change*, New York: Norton

Mansfield, E. (1983), 'Technological Change and Market Structure', *American Economic Review*, **73**, 205-209

Mansfield, E. (1984), 'R&D and Innovation: Some Empirical Findings', in Z. Griliches (ed.), *R&D, Patents, and Productivity*, Chicago, IL: University of Chicago Press for the National Bureau of Economic Research

Marshall, A. (1920), *Principles of Economics*, 8th edition, London: Macmillan

Martin, Roger (2007), *The Opposable Mind: How Successful Leaders Win Through Integrative Thinking*, Boston, MA: Harvard Business School Press

Marx, K. (1867/1974), *Capital: A Critical Analysis of Capitalist Production*, Volume 1, London: Lawrence and Wishart

Marx, K. and F. Engels (1848), *Manifesto of the Communist Party*, available at: http://www.efm.bris.ac.uk/het/marx/

Menge, J.A. (1962), 'Style Change Costs as a Market Weapon', *Quarterly Journal of Economics*, **76**, 632-647

Mensch, G. (1979), *Stalemate in Technology: Innovations Overcome the Depression*, Cambridge, MA: Ballinger

Merton, R.K. (1973), *The Sociology of Science*, Chicago, IL: Chicago University Press

Metcalfe, J.S. (1986), 'Information and Some Economics of the Information Revolution', in M. Ferguson (ed.), *New Communication Technologies and the Public Interest*, London: Sage

Mill, J.S. (1848/1923), *Principles of Political Economy*, London: Longmans and Co.

Mill, J.S. (1859/1929), *On Liberty*, London: Watts and Co.

Mowery, D. and N. Rosenberg (1979), 'The Influence of Market Demand upon Innovation: A Critical Review of Some Recent Empirical Studies', *Research Policy*, **8**, 102-153

Mueller, D.C. and Tilton, J.E. (1969), 'Research and Development Costs as a Barrier to Entry', *Canadian Journal of Economics* **2**, 570-579

Mumford, L. (1934), *Technics and Civilization*, New York: Harcourt Brace and Company

Mumford, L. (1961), *The City in History*, New York: Harcourt Brace and Company

Nelson, R.R. and S.G. Winter (1978), 'Forces Generating and Limiting Concentration under Schumpeterian Competition', *Bell Journal of Economics*, **9**, 524-548

Nelson R.R. and S.G. Winter (1982), *An Evolutionary Theory of Economic Change*, Cambridge, US: The Belknap Press of Harvard University

Nelson, R.R., S.G. Winter and H.L. Schuette (1976), 'Technical Change in an Evolutionary Model', *Quarterly Journal of Economics*, **90**, 90-118

Ohlin, B. (1933), *Interregional and International Trade*, Cambridge, MA: Harvard University Press

The Oxford Dictionary of Phrase, Saying and Quotation (1997), Oxford: Oxford University Press

Pavitt, K. (1987), 'On the Nature of Technology', Science Policy Research Unit Discussion Paper, University of Sussex, June

Pavitt, K. and S. Wald (1971), *The Conditions for Success in Technological Innovation*, Paris: OECD

Phillips, A. (1956), 'Concentration, Scale and Technological Change in Selected Manufacturing Industries, 1899-1939', *Journal of Industrial Economics*, **5**, 179-193

Phillips, A. (1966), 'Patents, Potential Competition and Technical Progress', *American Economic Review*, **56**, 301-310

Phillips, A. (1971), *Technology and Market Structure: A Study of the Aircraft Industry*, Lexington, MA: Heath/Lexington Books

Phlips, L. (1983), *The Economics of Price Discrimination*, Cambridge: Cambridge University Press

Porter, M. (1980), *Competitive Strategy*, New York: Free Press

Porter, M. (1990), *The Competitive Advantage of Nations*, London: Macmillan

Posner, M.V. (1961), 'International Trade and Technical Change', *Oxford Economic Papers*, **13**, 323-341

Rae, J. (1834), *Statement of Some New Principles on the Subject of Political Economy*, available at: http://www.efm.bris.ac.uk/het/rae/

Reinganum, J.F. (1985), 'Innovation and Industry Evolution', *Quarterly Journal of Economics*, **99**, 81-99

Rogers, E. (1995), *Diffusion of Innovations*, 4th edition, New York: Free Press

Rohlfs, J.H. (1974), 'A Theory of Interdependent Demand for a Communication Service', *Bell Journal of Economics and Management Science*, **5**, 16-37

Rubinstein, J. (1992), *The Changing U.S. Auto Industry: A Geographical Analysis*, London: Routledge

Ruskin, J. (1904/1996a), *The Stones of Venice*, in E.T. Cook and A. Wedderburn (eds), *The Works of John Ruskin*, Volume 10, CD-ROM Version, Cambridge: Cambridge University Press

Ruskin, J. (1905/1996b), *The Two Paths*, in E.T. Cook and A. Wedderburn (eds), *The Works of John Ruskin*, Volume 16, CD-ROM Version, Cambridge: Cambridge University Press

Ruskin, J. (1905/1996c), *Unto This Last*, in E.T. Cook and A. Wedderburn (eds), *The Works of John Ruskin*, Volume 17, CD-ROM Version, Cambridge: Cambridge University Press

Ruskin, J. (1905/1996d), *Munera Pulveris*, in E.T. Cook and A. Wedderburn (eds), *The Works of John Ruskin*, Volume 17, CD-ROM Version, Cambridge: Cambridge University Press

Salop, S. (1977), 'The Noisy Monopolist: Imperfect Information, Price Dispersion, and Price Discrimination', *Review of Economic Studies*, **44**, 393-406

Salop, S. and D. Scheffman (1983), 'Raising Rivals Costs', *American Economic Review*, **73**, 267-271

Saxenian, A. (1994), *Regional Advantage: Culture and Competition in Silicon Valley and Route 128*, Cambridge, MA: Harvard University Press

Scherer, F.M. (1979), 'The Welfare Economics of Product Variety: An Application to the Ready-to-Eat Breakfast Cereals Industry', *Journal of Industrial Economics*, **28**, 113-134

Scherer, F.M. (1984), *Innovation and Growth: Schumpeterian Perspectives*, Cambridge, MA: MIT Press

Schmalensee, R. (1978), 'Entry Deterrence in the Ready-to-Eat Breakfast Cereal Industry', *Bell Journal of Economics*, **9**, 305-327

Schmalensee, R. (1982), 'Product Differentiation Advantages of Pioneering Brands', *American Economic Review*, **72**, 349-365

Schumacher, E. (1974), *Small is Beautiful*, London: Abacus/Sphere Books

Schumpeter, J.A. (1954), *Capitalism, Socialism and Democracy*, 4th edition, London: Unwin University Books

Scitovsky, T. (1976), *The Joyless Economy: An Enquiry into Human Satisfaction and Consumer Dissatisfaction*, Oxford; Oxford University Press

Senior, N. (1863), *Outline of the Science of Political Economy*, 5th edition, London: Charles Griffin

Sheldon, K. M. (1999), 'Conformity', in M.A. Runco and S.R. Pritzker (eds), *Encyclopaedia of Creativity*, Volume 1, San Diego, CA: Academic Press, pp. 341-346

Shepherd, W.G. (1990), *The Economics of Industrial Organisation*, 3rd edition, Englewood Cliffs, NJ: Prentice Hall International

Simon, H.A. (1985), 'What Do We Know about the Creative Process?', in R.L. Kuhn (ed.), *Frontiers in Creative and Innovative Management*, Cambridge, MA: Ballinger

Singh, A. and G. Whittington (1975), 'The Size and Growth of Firms', *Review of Economic Studies*, **42**, 15-26

Smith, A. (1776/1904a), *An Inquiry into the Nature and Causes of the Wealth of Nations,* Volume I, London: Methuen

Smith, A. (1776/1904b), *An Inquiry into the Nature and Causes of the Wealth of Nations,* Volume II, London: Methuen

Solow, R.M. (1957), 'Technical Change and the Aggregate Production Function', *The Review of Economics and Statistics*, **39**, 312-320

Solow, R.M. (1987), 'We'd Better Watch Out', *New York Times*, 12 July

Stamm, B. von (2003), *Managing Innovation, Design and Creativity*, Chichester, UK: John Wiley and Sons

Stern Review on the Economics of Climate Change (2006), available at http://www.hm-treasury.gov.uk/sternreview_index.htm

Stonebraker, R.J. (1976), 'Corporate Profits and the Risk of Entry', *Review of Economics and Statistics*, **58**, 33-39

Stoneman, P. (1987), *The Economic Analysis of Technology Policy*, Oxford: Oxford University Press

Swann, G.M.P. (1990), 'Product Competition and the Dimensions of Product Space', *International Journal of Industrial Organisation*, **8**, 281-295

Swann, G.M.P. (2001a), 'The Demand for Distinction and the Evolution of the Prestige Car', *Journal of Evolutionary Economics*, **11**, 59-75

Swann, G.M.P. (2001b), 'Will the Internet Lead to Perfect Competition?', *The Business Economist*, **32**, 6-14

Swann, G.M.P. (2002a), 'The Functional Form of Network Externalities', *Information Economics and Policy*, **14**, 417-429

Swann, G.M.P. (2002b), 'There's More to the Economics of Consumption than (Almost) Unrestricted Utility Maximisation', in A. McMeekin, K. Green, M. Tomlinson and V. Walsh (eds), *Innovation By Demand: An Interdisciplinary Approach to the Study of Demand and its Role in Innovation*, Manchester: Manchester University Press, pp. 23-40

Swann, G.M.P. (2010, forthcoming), *Prosperity through Innovation: How we Create the Wealth of Nations*, Cheltenham, UK and Northampton, MA, USA: Edward Elgar

Swann, G.M.P. and J. Gill (1993), *Corporate Vision and Rapid Technological Change*, London: Routledge

Swann, G.M.P., M. Prevezer and D. Stout (eds) (1998), *The Dynamics of Industrial Clustering*, Oxford: Oxford University Press

Swann, G.M.P. and M. Taghavi (1992), *Measuring Price and Quality Competitiveness*, Aldershot, UK: Avebury

Symonds, R.W. (1947), *A History of English Clocks*, Harmondsworth, UK: Penguin Books

Taylor, P., J. Beaverstock, G. Cook, N. Pandit, K. Pain, H. Greenwood (2003), *Financial Services Clustering and its Significance for London*, Corporation of London, http://www.cityoflondon.gov.uk/Corporation/

Teece, D., G. Pisano and A. Shuen (1991), 'Firm Capabilities, Resources and the Concept of Strategy', CCC Working Paper no. 90-8 (Revised), University of California at Berkeley, Center for Research in Management

Temple, P., K. Blind, A. Jungmittag, C. Spencer, G.M.P. Swann and R. Witt (2005), *The Empirical Economics of Standards*, DTI Economics Paper no. 12, London: Department of Trade and Industry, available at: http://www.berr.gov.uk/files/file9655.pdf

Tirole, J. (1988), *The Theory of Industrial Organization*, Cambridge, MA: MIT Press

Tushman, M. and P. Anderson (1986), 'Technological Discontinuities and Organisational Environments', *Administrative Science Quarterly* **31**, 439-465

Urban, G., T. Carter, S. Gaskin and Z. Mucha (1984), 'Market Share Rewards to Pioneering Brands', *Management Science*, **32**, 645-659

Varian, H. (2000), 'Priceline's Magic Show', *The Industry Standard Magazine*, 24 April

Veblen, T. (1899), *The Theory of the Leisure Class: An Economic Study of Institutions,* New York: Macmillan

Veblen, T. (1914), *The Instinct of Workmanship and the State of the Industrial Arts*, New York: Macmillan

Vernon, R. (1966), 'International Investment and International Trade in the Product Cycle', *Quarterly Journal of Economics*, **80**, 190-207

Waldrop, M.M. (1994), *Complexity: The Emerging Science at the Edge of Order and Chaos*, Harmondsworth, UK: Penguin Books

Warde, A. (2002), 'Social Mechanisms Generating Demand: A Review and Manifesto', in A. McMeekin, K. Green, M. Tomlinson and V. Walsh (eds), *Innovation By Demand: An Interdisciplinary Approach to the Study*

of Demand and its Role in Innovation, Manchester: Manchester University
Press, pp. 10-22

Winter, S.G. (1984), 'Schumpeterian Competition in Alternative
Technological Regimes', *Journal of Economic Behaviour and
Organisation*, **5**, 287-320

Wyatt, G. (1986), *The Economics of Invention*, Brighton, UK: Wheatsheaf
Books

Yoffie, D.B. and M.A. Cusumano (1999), 'Judo Strategy: The Competitive
Dynamics of Internet Time', *Harvard Business Review*, **77**, 70-81

Websites referred to:

http://en.wikipedia.org/wiki/Burton_upon_Trent_brewing
http://en.wikipedia.org/wiki/Open_source
http://news.ft.com/
http://www.berr.gov.uk/
http://www.berr.gov.uk/dius/innovation/
http://www.berr.gov.uk/dius/innovation/randd
http://www.berr.gov.uk/sectors/biotech/
http://www.businessweek.com/
http://www.camra.org.uk/
http://www.chemicalsnw.org.uk/
http://www.cityoflondon.gov.uk/Corporation/
http://www.dius.gov.uk/
http://www.diva-portal.org/kth/
http://www.efm.bris.ac.uk/het/
http://www.guardian.co.uk
http://www.hm-treasury.gov.uk/gowers_review_index.htm
http://www.hm-treasury.gov.uk/sternreview_index.htm
http://www.innovationindex.org.uk/
http://www.ipo.gov.uk/
http://www.london.gov.uk/gla/publications/
http://www.opensource.org/
http://www.priceline.com/
http://www.seeda.co.uk/
http://www.sqw.co.uk/
http://www.thepotteries.org/

Index of names

Index of topics